T0365993

Sustainable Manufacturing

Sustainable Manufacturing

Concepts, Tools, Methods and Case Studies

S. Vinodh

CRC Press is an imprint of the
Taylor & Francis Group, an **informa** business

First edition published 2020
by CRC Press
6000 Broken Sound Parkway NW, Suite 300, Boca Raton, FL 33487-2742

and by CRC Press
2 Park Square, Milton Park, Abingdon, Oxon, OX14 4RN

Library of Congress Cataloging-in-Publication Data

Names: Vinodh, S., author.
Title: Sustainable manufacturing : concepts, tools, methods and case studies / S. Vinodh.
Description: First edition. | Boca Raton : CRC Press, 2020. | Includes bibliographical references and index.
Identifiers: LCCN 2020020935 | ISBN 9780367278687 (hbk) | ISBN 9780429320842 (ebk)
Subjects: LCSH: Lean manufacturing. | Sustainable engineering. | Manufacturing processes--Environmental aspects.
Classification: LCC TS155.7 .V56 2020 | DDC 628--dc23
LC record available at https://lccn.loc.gov/2020020935

ISBN: 9780367278687 (hbk)
ISBN: 9780429320842 (ebk)

Typeset in Palatino
by Deanta Global Publishing Services, Chennai, India

This book is dedicated to my parents, aunt, wife, son, teachers, professors, students, friends and well-wishers.

Contents

Preface xiii
Acknowledgements xv
Author xvii

1. Concepts and Fundamentals of Sustainable Manufacturing, Triple Bottom Line Approach 1
1.1 Overview of Triple Bottom Line 1
1.2 Overview of Sustainable Manufacturing with Evolutionary Aspects 1
1.2.1 Evolution of Sustainable Manufacturing 3
1.3 Definitions of Sustainable Manufacturing 3
1.3.1 Business Benefits of SM are (Alayon, 2016) 4
1.3.2 Implementation of Sustainable Manufacturing at Initial Stage 4
1.4 Linkage between Lean and Sustainable Manufacturing 4
1.4.1 Basic Lean Waste and Corresponding Environmental Aspects 5
1.4.2 Influence of Lean Manufacturing on Three Dimensions of Sustainability 5
1.5 Summary 6
References 6

2. Sustainable Manufacturing Tools: Environmentally Conscious Quality Function Deployment (ECQFD) 9
2.1 ECQFD Phase I 9
2.2 ECQFD Phase II 10
2.3 ECQFD Phase III 10
2.4 ECQFD Phase IV 11
2.5 Extended Producer Responsibility (EPR) Policy 13
2.5.1 Unique Features of EPR Policy 13
2.5.2 Potential Benefits of EPR Policy 13
2.6 Summary 13
References 14

3. Sustainable Manufacturing Tools: Life Cycle Assessment (LCA) 15
3.1 Fundamentals of LCA 15
3.1.1 Applications of LCA 15
3.1.2 Steps of LCA 15
3.2 LCA Phase I: Goal and Scope Definition 16
3.2.1 Functional Unit 16
3.2.2 System Boundaries 16

3.3 LCA Phase II: Inventory Analysis ... 17
3.3.1 Types of Data Required in LCA ... 17
3.4 LCA Phase III: Impact Assessment ... 17
3.4.1 Classification ... 17
3.4.2 Characterization ... 18
3.4.3 Normalization ... 18
3.4.4 Grouping ... 18
3.4.5 Weighting ... 18
3.5 LCA Phase IV: Interpretation ... 18
3.5.1 Evaluation ... 19
3.5.2 Conclusions, Recommendations and Reporting ... 19
3.5.3 Report ... 19
3.6 Life Cycle Cost Analysis ... 19
3.7 Summary ... 21
References ... 21

4. Life Cycle Impact Assessment Methods: CML, Eco-Indicator 95 and Eco-Indicator 99 ... 23
4.1 CML Method ... 23
4.1.1 Description ... 23
4.1.2 Normalization ... 24
4.1.3 Weighting ... 24
4.1.4 Features of CML ... 24
4.1.5 Quality Considerations ... 24
4.2 Eco-Indicator Method ... 25
4.2.1 Principles of the Eco-Indicator ... 25
4.2.2 Scope of Eco-Indicator ... 25
4.2.3 Environmental Effects ... 25
4.2.4 Operating Instructions ... 25
4.2.5 Description of the Eco-Indicator 95 Method ... 26
4.3 Eco-Indicator 99 Method ... 26
4.3.1 Different Archetypes in Eco-Indicator 99 ... 27
4.3.2 Features of Eco-Indicator 99 ... 27
4.4 ReCiPe Method ... 28
4.4.1 Overview ... 28
4.4.2 Connectivity between Midpoint and Endpoint Levels ... 28
4.4.3 Dealing with Uncertainties and Assumptions: Scenarios ... 28
4.4.4 Unique Features of ReCiPe 2008 ... 29
4.4.5 Normalization in ReCiPe 2008 ... 29
4.4.6 Weighting in ReCiPe 2008 ... 29
4.5 Summary ... 29
References ... 29

5. Design Strategies Supporting Sustainable Manufacturing 31
5.1 Design for Disassembly 31
5.1.1 Principles 31
5.1.2 Scope 31
5.1.3 Need 31
5.1.4 Benefits 32
5.1.5 Factors Affecting Disassembly 32
5.1.6 Types of Disassembly 32
5.1.7 Designing for Active Disassembly 33
5.2 Design for Recycling 33
5.2.1 Scope 34
5.2.2 Need 34
5.2.3 Recycling Process 34
5.2.4 Types of Recycling 34
5.2.5 Benefits 34
5.3 Design for Environment (DFE) 34
5.3.1 Need for DFE 35
5.3.2 Practices 35
5.3.3 Benefits 35
5.3.4 Eco Design Strategy Wheel 36
5.3.5 DFE Process 36
5.4 Eco-Friendly Product Design Methods 37
5.4.1 Definition 37
5.4.2 Scope 37
5.4.3 Need 37
5.4.4 Tools That Include Environmental Needs in the Design Process 38
5.4.4.1 Tools Based on Design Matrix 38
5.4.4.2 Tools Based on Quality Function Deployment 38
5.4.4.3 Tools Based on Value Analysis (VA) 38
5.4.4.4 Tools Based on Failure Mode and Effect Analysis (FMEA) 38
5.4.4.5 Other Tools 39
5.4.5 Review of Eco Design Strategies and Methodologies 39
5.4.6 Steps Involved in Eco-Innovative Product Design 39
5.4.7 Benefits of Eco Design Strategies 39
5.5 Summary 40
References 40

6. Standards for Sustainable Manufacturing 43
6.1 ISO 14001 Environmental Management Systems (EMS) 43
6.1.1 Clauses of ISO 14001 43
6.1.2 Advantages of Implementing ISO 14000 45

6.1.3 Certain ISO 14000 Series Standards for sustainability (Source: ISO 2002) (Christini et al., 2004). 45
6.1.4 Disadvantages of EMS 45
6.2 PAS 2050 Standard 45
6.2.1 Need for PAS 2050 46
6.2.2 Phases of PAS 2050 46
6.2.3 Organizational Benefits 46
6.2.4 Concept of Carbon Footprint 46
6.3 Summary 47
References 47

7. Product Sustainability and Risk–Benefit Assessment, and Corporate Social Responsibility 49
7.1 Product Sustainability and Risk–Benefit Assessment 49
7.2 Risk–Benefit Assessment Case 50
7.3 Corporate Social Responsibility (CSR): Overview 50
7.3.1 Definitions 54
7.3.2 Scope 54
7.3.3 Need for CSR are (Sprinkle & Maines, 2010) 55
7.3.4 Benefits of CSR are (Sprinkle & Maines, 2010) 55
7.3.4.1 Business Benefits of CSR are (Drews, 2010) 55
7.4 Drivers of CSR 55
7.5 Summary 55
References 56

8. Sustainability Assessment 59
8.1 Sustainability Indicators (Environment, Economy and Society Based) 59
8.2 Sustainability Assessment Models 59
8.2.1 A Generic Framework for Sustainability Assessment of Manufacturing Processes 59
8.2.2 Integrated Sustainability Assessment Framework 60
8.2.3 Three-Level Conceptual Model for Sustainability Assessment 61
8.2.4 Sustainability Assessment Model for SMEs 62
8.2.5 A Metrics-Based Framework to Assess Total Life Cycle Sustainability of Manufactured Products 62
8.2.6 Sustainability Assessment Using Fuzzy Inference Technique 62
8.2.7 Sustainability Evaluation Model for Manufacturing Systems 62
8.3 Multi-Grade Fuzzy Assessment Approach 63
8.4 Fuzzy Logic Assessment Approach 64
8.5 Summary 67
References 67

9. Software Modules for Life Cycle Assessment (LCA) and Sustainable Manufacturing 69
9.1 Product-Based Sustainability Analysis Module 69
9.1.1 Sustainability Xpress: SolidWorks Sustainability Analysis Module 69
9.1.2 Need 69
9.1.3 Scope 69
9.1.4 Environmental Impact 70
9.1.5 Steps in Sustainability Xpress 70
9.1.6 Sustainability Report 71
9.1.6.1 Analysis Results of a Case Study Product 71
9.2 Process-Based LCA–GaBi Module 72
9.2.1 Features of GaBi are (Spatari et al., 2001; PE-International, 2012) 73
9.2.2 Modelling and Analysis Using GaBi 74
9.2.2.1 Goal and Scope Definition 74
9.2.2.2 Life Cycle Inventory (LCI) 74
9.2.2.3 Impact Assessment 75
9.2.2.4 Interpretation 75
9.2.3 Procedural Steps in GaBi 75
9.2.4 Case Study 75
9.3 Process-Based LCA–SimaPro Module 76
9.3.1 Features of SimaPro Module are (Goedkoop et al., 2016) 78
9.3.2 Steps in SimaPro Module 78
9.3.3 Case Study 79
9.4 Summary 80
References 83

10. Sustainability and Energy Aspects of Manufacturing Processes 85
10.1 Sustainability of Conventional Manufacturing Processes 85
10.1.1 Framework-Based Studies 85
10.1.2 Energy-Based Studies 85
10.1.3 Life Cycle Assessment (LCA)-Based Studies 86
10.1.4 Inferences from Certain Research Studies 86
10.2 Sustainability of Unconventional Manufacturing Processes 86
10.2.1 Sustainability Aspects of ECM, EDM and USM Processes 86
10.2.2 LCA-Based Studies 87
10.3 Sustainability of Additive Manufacturing Processes 87
10.3.1 LCA Studies of AM 87
10.3.2 Energy Evaluation in AM 88
10.3.3 Design for Additive Manufacturing (DFAM) Guidelines 89
10.4 Summary 90
References 90

11. Case Studies on ECQFD, LCA and MCDM 93
11.1 Case on Environmentally Conscious Quality Function Deployment (ECQFD) 93
11.2 Case on Life Cycle Costing and Life Cycle Assessment 102
11.2.1 Product Life Cycle Costing 102
11.2.2 Case Study on LCA 102
11.3 Case on MCDM 105
11.4 Summary 113

12. Research Issues in Sustainable Manufacturing 115
12.1 Relation between Lean and Sustainable Manufacturing 115
12.2 Multi-Criteria Decision Making (MCDM) and Easily Implemented MCDM methods in Sustainable Manufacturing 115
12.3 Research Agenda in Sustainable Manufacturing 116
12.4 Sustainable Manufacturing Portal 117
12.5 Sustainable Manufacturing for Industry 4.0 118
12.6 Summary 118
References 119

Index 121

Preface

Modern manufacturing organizations have been adopting sustainable manufacturing to develop environmentally friendlier products. Sustainable manufacturing enables the development of products with minimal environmental impact coupled with economic and societal benefits. Sustainable manufacturing is governed by the triple bottom line approach focusing on environmental, economic and societal dimensions. The proposed book narrates the concepts, tools/techniques and design strategies supporting sustainable manufacturing.

Salient topics include:

- Tools/techniques of sustainable manufacturing
- Standards of sustainable manufacturing
- Design strategies supporting sustainable manufacturing
- Performance measures of sustainable manufacturing
- Case studies on sustainable manufacturing
- Software modules for life cycle assessment and sustainable manufacturing
- Research insights on sustainable manufacturing

The book provides descriptions of concepts, supporting illustrations and examples. This book focuses on the fundamentals and research aspects of sustainable manufacturing. Case studies on sustainable manufacturing are presented to depict the practical perspectives of sustainable manufacturing.

It is believed that the book enables readers to understand the concepts, tools/techniques, standards and performance measures of sustainable manufacturing. Case studies on sustainable manufacturing are illustrated with research highlights.

Chapter 1 is an introduction to sustainable manufacturing with an overview of the triple bottom line approach and sustainable manufacturing with evolutionary aspects. The chapter also reviews key definitions of sustainable manufacturing, and explores the linkage between lean and sustainable manufacturing.

Chapter 2 gives an idea of the sustainable manufacturing tool environmentally conscious quality function deployment (ECQFD) wherein four phases are detailed with the extended producer responsibility (EPR) policy.

Chapter 3 details the fundamentals of a life cycle assessment (LCA), activities in the four phases of LCA and life cycle cost analysis.

Chapter 4 provides details on life cycle impact assessment (LCIA) methods: CML, Eco-indicator 95 and 99; and ReCiPe.

Chapter 5 is devoted to design strategies enabling sustainable manufacturing: design for disassembly, design for recycling, design for environment and eco-friendly product design.

Chapter 6 explains standards for sustainable manufacturing: ISO 14000 Environmental Management System and PAS 2050.

Chapter 7 explains product sustainability and risk–benefit assessment, as well as corporate social responsibility along with its drivers.

Chapter 8 provides details on sustainability indicators, sustainability assessment models and details of sustainability assessment approaches.

Chapter 9 is devoted to software modules of LCA and sustainable manufacturing wherein product-based LCA and process-based LCA modules are illustrated.

Chapter 10 provides details on sustainability and energy aspects of conventional, unconventional and additive manufacturing processes.

Chapter 11 details case studies on environmentally conscious quality function deployment, life cycle costing and life cycle assessment, and multi-criteria decision making.

Chapter 12 is dedicated to research issues in sustainable manufacturing wherein the relationship between lean and sustainable manufacturing, multi-criteria decision making in sustainable manufacturing, the sustainable manufacturing portal and sustainable manufacturing for Industry 4.0 are detailed.

Finally I would like to thank all readers who will be using this book and I wish you the best of luck with your endeavours.

Acknowledgements

The motivation for the development of this book originates from my research work and publications over the past ten years.

I sincerely thank Almighty God for providing energy and strength to complete writing the book.

I sincerely thank the director and administration of our institute and my department for providing the necessary infrastructure and support for book writing.

I thank my parents, aunt, wife, son, sister, nephew and other family members for their care and moral support rendered during book writing.

I wholeheartedly thank my scholar Rohit for his continued support during various stages of book writing.

I thank my professors, friends and well-wishers for their motivation. My special thanks to my beloved professor and his mother for their motivation and support during book writing.

I thank all my research group members (past and present students) for their support in various research studies that got published in international journals, which formed the foundation of this book.

Finally, I would like to thank the publisher CRC Press (Taylor & Francis Group) and its editorial team for their help and support during various stages of book publication.

Author

S. Vinodh is an Associate Professor in the Department of Production Engineering, National Institute of Technology, Tiruchirappalli, Tamil Nadu, India. He completed his PhD under the AICTE National Doctoral Fellowship scheme from PSG College of Technology, Coimbatore, India. He completed his master's degree in production engineering from PSG College of Technology and bachelor's degree in mechanical engineering from Government College of Technology, Coimbatore, India. He was a Gold Medallist in his undergraduate study. He has published over 150 papers in international journals. He has about ten years of research experience in sustainable manufacturing. He received the Highly Commended Paper Award from Emerald Publishers for the year 2016. He is the recipient of the Institution of Engineers Young Engineer Award for the year 2013 in the Production Engineering Division. He is the recipient of the Innovative Student Project Award 2010 based on his PhD thesis from the Indian National Academy of Engineering (INAE), New Delhi, India. He has executed research projects and guided PhD scholars in sustainable manufacturing. His research interests include sustainable manufacturing, lean manufacturing, agile manufacturing, rapid manufacturing and Product Development and Industry 4.0.

1

Concepts and Fundamentals of Sustainable Manufacturing, Triple Bottom Line Approach

1.1 Overview of Triple Bottom Line

Triple bottom line (TBL) is the base for sustainability. TBL underlies the rationale of sustainability by evaluating the impact in terms of profitability; shareholder values; and its societal, people and environmental capital (Savitz, 2013). It is based on 3Ps: profitability, people and planet. Profitability pertains to the economy; people pertain to societal and planet pertains to environmental dimensions. It considers the development and application of a corporate strategy involving environmental, societal and financial results (Mowat, 2002). TBL provides a three-dimensional view of sustainability moving from the viewpoint of the environment to include economic and societal dimensions (Shaffer, 2018). Widely cited definitions of TBL are presented in Table 1.1.

1.2 Overview of Sustainable Manufacturing with Evolutionary Aspects

Manufacturing systems have witnessed a shift from mass production to lean, green and sustainable manufacturing. The transition of manufacturing systems is shown in Figure 1.1. Key parameters governing this transition include product complexity, market dynamism and stakeholder value. Craft manufacturing includes skilled employees and has low production volume. Mass manufacturing is characterized by interchangeable parts and delivers high productivity (Rojko, 2017). It can handle market dynamism better than a craft system. Mass manufacturing is characterized by high-volume manufacturing with limited product variants. It fulfils economies of scale, which state that the unit cost of a product comes down as a result of high-volume production. Lean manufacturing is characterized by waste elimination,

TABLE 1.1
Widely Cited Definitions of TBL

Definition	Reference
'Taking environmental, social and financial results into consideration in the development and implementation of a corporate business strategy'.	Mowat (2002)
TBL firms aim to become more responsive ecologically and socially while prospering economically. This threefold focus is often referred to in terms of 'people, planet, profit', or as the "triple bottom line' (TBL).	Elkington (1997)

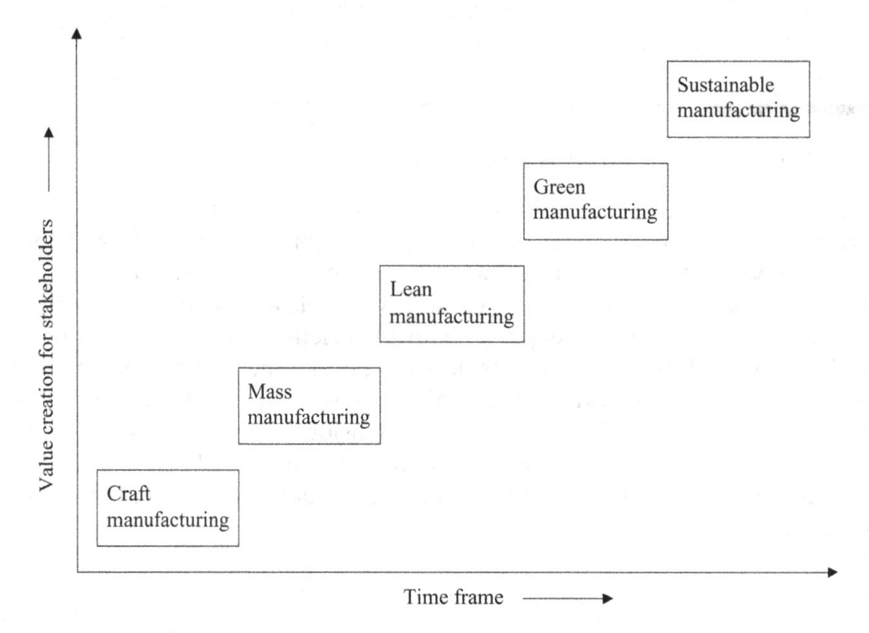

FIGURE 1.1
Transition of manufacturing systems.

streamlined processes and value addition (de Freitas et al., 2017). Lean is a predecessor manufacturing system to green and sustainable manufacturing. Lean production is based on the Toyota Production System (TPS), which is waste reduction based on skilled workers and focused on value addition. Green manufacturing is environmentally benign and 3R-based (reduce, reuse, recycle). Sustainable manufacturing is innovative and 6R-based (3Rs plus recovery, redesign and remanufacture). Sustainable manufacturing is referred to as an 'E' paradigm with the focus on ecology, economy, excellence and so on. Lean is a predecessor element to sustainability as lean facilitates waste elimination, which forms the path for sustainable manufacturing.

Sustainable manufacturing (SM) is concerned with executing a new form of business with value creation to develop green products and processes in demand (Alayón, 2016). Key business benefits from SM include improvement of efficiency and productivity, reduced usage of hazardous materials, compliance with regulations, enhanced reputation and enhanced community relations.

The environmental focus of manufacturing was initially on pollution abatement (reducing pollution). Then the shift was towards cleaner production and pollution prevention (U.S. Department of Commerce, 2011).

1.2.1 Evolution of Sustainable Manufacturing

The main environmental focus in a manufacturing context had been referred to as pollution abatement, that is preventing the pollution generated from entering the environment. Then the focus became cleaner production and pollution prevention. With cleaner production, there exists scope for cost reduction and other economic advantages (U.S. Department of Commerce, 2011).

- Clean technologies are vital from the viewpoint of sustainable manufacturing and include associated aspects such as environmental protection, compliance with regulatory bodies, pollution prevention and renewable technologies.
- Green production deals with development of green products, i.e. products with minimal environmental impact.
- Sustainable manufacturing includes a methodological approach for waste elimination by resource optimization and utilization of resources, and technologies with less environmental impact.

1.3 Definitions of Sustainable Manufacturing

Widely cited definitions of sustainable manufacturing are presented in Table 1.2. Elements identified with sustainable manufacturing include fewer environmental impacts, resources conservation, safety, economic benefits,

TABLE 1.2

Widely Cited Definitions of Sustainable Manufacturing

Definition	Reference
'Manufacturing processes that meet the needs of the present without compromising future generations' ability to meet their own needs'.	Gardner and Colwill (2016)
'The creation of manufactured products that use processes that minimize negative environmental impacts, conserve energy and natural resources, are safe for employees, communities, and consumers; and are economically sound'.	U.S. Department of Commerce (2011)

meeting requirements of future generations, energy conservation, employee health, inclusive growth and optimal resource utilization. SM facilitates the development of green products that are designed to reduce environmental impact with the usage of recyclable materials.

1.3.1 Business Benefits of SM are (Alayon, 2016)

- Increased sales.
- Improved efficiency and productivity.
- Compliance with regulations.
- Enhanced reputation and image.
- Better community relations (Alayón, 2016).

1.3.2 Implementation of Sustainable Manufacturing at Initial Stage

The details of implementing SM are presented as follows (U.S. Department of Commerce, 2011; Kishawy et al., 2018):

- Housekeeping – Can be facilitated with 5S lean tools for enhancement in work practices.
- Process optimization – Deals with changes in manufacturing processes in terms of process conditions and parameters to enhance efficiency.
- Raw material substitution – The selection of alternative materials to deal with minimal environmental impact.
- New technologies – Exploring new technologies with less resource consumption and fewer emissions.
- New product design – A focus on designing products with minimal environmental impact across life cycle phases.

1.4 Linkage between Lean and Sustainable Manufacturing

Lean concepts facilitate workplace improvement thereby ensuring health and wellness of the workforce with organized layouts and minimized safety hazards. Waste reduction is a vital issue for attaining environmental sustainability. Lean practices facilitate less material usage, fewer scraps, and reduced water and energy utilization. Concurrent implementation of sustainability and lean practices enhance business performance with fulfilling benefits in environmental, economic and societal dimensions (de Freitas et al., 2017).

If the organization adopts lean practices, it would be better to amalgamate environmental principles into the lean process. In addition to the original seven wastes of lean with the eighth one (underutilization of workforce creativity), a ninth waste is environmental waste, through which lean techniques have scope for reducing environmental wastes. Environmental waste is defined as unwanted resources utilization or entities released into air, water or land that impact the environment. Environmental waste includes overutilization of energy, water or any other material; hazardous materials; and pollutants.

1.4.1 Basic Lean Waste and Corresponding Environmental Aspects

The fundamental lean wastes and their related environmental wastes are presented as follows (U.S. Department of Commerce, 2011; Choudhary et al., 2019):

- Unnecessary transport or motion – Energy used for transport that generates emissions. Extra space is required for extra motion.
- Inventory – Energy is required for heating, cooling and lighting of store space. Additional packing is required for additional stock.
- Waiting – Deals with materials spoilage or damaged materials during delays.
- Overproduction – The usage of materials to generate products (not needed) and their emissions.
- Processing – Unnecessary processing consumes resources and increases waste.
- Defects – Resources utilized to make defective parts, as well as their disposal.

1.4.2 Influence of Lean Manufacturing on Three Dimensions of Sustainability

The details of the three dimensions of sustainability are presented as follows (Varela et al., 2019):

- Economic dimension – Enhancement of profit and market share, reduced operational cost, enhanced process performance.
- Environment dimension – Reduced energy consumption, reduced industrial waste, adopting circular economy practice.
- Social dimension – Enhanced quality of work conditions, enhanced financial benefits, increased health and safety practices (Saurin et al., 2006; Koskela, 1993).

1.5 Summary

This chapter provides an overview of triple bottom line with widely cited definitions. Evolutionary aspects of sustainable manufacturing highlighting the transition from craft to mass manufacturing, followed by lean, green and sustainable manufacturing are highlighted. The characteristics of manufacturing systems are mentioned. Widely cited definitions of sustainable manufacturing are presented with key terms. The connectivity between lean and sustainability, and the influence of lean on triple bottom line sustainability dimensions are presented.

References

Alayón, C. (2016). *Exploring sustainable manufacturing principles and practices* (Doctoral dissertation). Jonkoping: Jönköping University, School of Engineering.

Choudhary, S., Nayak, R., Dora, M., Mishra, N., & Ghadge, A. (2019). An integrated lean and green approach for improving sustainability performance: A case study of a packaging manufacturing SME in the UK. *Production Planning and Control, 30*(5–6), 353–368.

de Freitas, J.G., Costa, H.G., & Ferraz, F.T. (2017). Impacts of Lean Six Sigma over organizational sustainability: A survey study. *Journal of Cleaner Production, 156*, 262–275.

Elkington, J. (1997). *Cannibals with forks: The triple bottom line of twenty-first century business.* Oxford, UK: Capstone Press.

Gardner, L., & Colwill, J. (2016). A framework for the resilient use of critical materials in sustainable manufacturing systems. *Procedia CIRP, 41*, 282–288.

Kishawy, H.A., Hegab, H., & Saad, E. (2018). Design for sustainable manufacturing: Approach, implementation, and assessment. *Sustainability, 10*(10), 3604.

Koskela, L. (1993). Lean production in construction. *Automation in Construction, 3*(1), 47–54.

Mowat, D. (2002). The VanCity difference: A case for the triple bottom line approach to business. *Corporate Environmental Strategy, 9*(1), 24–29.

Rojko, A. (2017). Industry 4.0 concept: Background and overview. *International Journal of Interactive Mobile Technologies, 11*(5), 77–90.

Saurin, T.A., Formoso, C.T., & Cambraia, F.B. (2006). Towards a common language between lean production and safety management. *Proceedings of the 14th annual conference of the international group for lean construction (IGLC-14).* Ketchum, ID: Lean Construction Institute.

Savitz, A. (2013). *The triple bottom line: How today's best-run companies are achieving economic, social and environmental success: And how you can too.* Chichester: John Wiley & Sons.

Shaffer, G.L. (2018). *Creating the sustainable public library: The triple bottom line approach.* Santa Barbara, CA: ABC-CLIO.

U.S. Department of Commerce. (2011). *Introduction to sustainable manufacturing.* Washington, DC: International Trade Administration Manufacturing and Services.

Varela, L., Araújo, A., Ávila, P., Castro, H., & Putnik, G. (2019). Evaluation of the relation between lean Manufacturing, Industry 4.0, and Sustainability. *Sustainability, 11*(5), 1439.

2

Sustainable Manufacturing Tools: Environmentally Conscious Quality Function Deployment (ECQFD)

Environmentally conscious quality function deployment (ECQFD) is the modified form of quality function deployment (QFD) which can be applied to environmental and sustainability-based studies.

The procedure of ECQFD is as follows (Masui et al., 2003):

- Identify voice of customer (VOC) and engineering metrics (EM)

  This step includes the identification of customer requirements from the environmental viewpoint which needs to be included in product development and integrating those customer requirements into the product life cycle. The voice of customer (VOC) is then converted into engineering terms, also called engineering metrics (EMs). The added advantage of ECQFD over QFD is that it includes a third dimension in the QFD matrix: environmental aspects.
- Identify environmental VOCs

  Environmental VOCs are the customer requirements which must be included in the product life cycle in the design phase itself to solve environment-related problems.
- Identify environmental EMs

  Environmental EMs include technical properties of the product which need to be considered to ensure VOC.
- Identify opportunities for design enhancement

  This includes phases (I and II) of ECQFD.

2.1 ECQFD Phase I

Phase I depicts the ECQFD matrix integrating VOCs and EMs. VOCs are written in rows and EMs are written in columns. The weights of VOCs are then specified based on expert opinion. The customer weights are given based on their importance. A rating of 9 indicates 'more importance', 3 indicates 'medium importance' and 1 indicates 'less importance'. Then mapping needs to be done among the VOCs and EMs; this is also called the relational strength between a VOC and EM. A rating between a VOC and EM determines its relational strength. A rating of 9 denotes 'strong relationship', 3 denotes 'moderate relationship' and 1 denotes 'low relationship'. This relational strength helps designers in decision making. After assigning weights and ratings to VOCs and EMs, the next step is to calculate the raw score. The raw score for each EM is the sum of the product of customer weight and the relational strength. Then the relative weight of each EM needs to be calculated. The relative weight of each EM is the ratio of the raw score of the EM to the total raw score of all EMs (Masui et al., 2003).

2.2 ECQFD Phase II

Phase II includes the integration of EMs in product components. In phase II, EMs are written in rows and product components are written in columns. The relative weight of each EM is taken from phase I. The relative weight of each part is calculated similar to phase I.

- Evaluate considered design improvements
  It includes phases III and IV of ECQFD.

2.3 ECQFD Phase III

In phase III, the focus is to evaluate the impact of design improvements on the EM. In this phase, the designer can make several alternative options for improvement. The options can be made in two different ways. One way is to fix the target VOC and the second method is to evaluate important components from phase II. The two design options include different combinations of EM and component. In ECQFD, the main focus is given to the environment and the improvement plan needs to be fixed by considering environmental aspects. The improvement rate of EMs need to be evaluated in this phase. The equation for calculating improvement rate of EMs is represented as (Masui et al., 2003; Vinodh & Rathod, 2011)

$$\text{Improvement rate of engineering metrics } \left(IR_j\right) = \frac{\sum_{K=1}^{K}\left(r_{j,k} * c_{j,k}\right)}{\sum_{K=1}^{K}\left(r_{j,k}\right)} \tag{2.1}$$

where

IR_j is the improvement rate of engineering metrics EM_j.

$r_{j,k}$ is the relational intensity between component C_k and engineering metrics EM_j.

$C_{j,k}$ is the improvement rate of engineering metrics EM_j with respect to component C_k; C_k has a value of either 0 or 1 depending on whether improvement is possible or not possible.

2.4 ECQFD Phase IV

In phase IV, the impact of design improvement on EMs into customer requirements is assessed. The phase IV matrix includes VOCs in rows and EMs in columns. The customer weight and the relational strength in phase IV is the same as that of phase I. The improvement rates of EMs are taken from phase III. The improvement rate (IR) of customer requirement (CR) needs to be calculated in this phase.

The improvement rate of customer requirement (IR_i) can be calculated using Equation 2.2 (Masui et al., 2003; Vinodh & Rathod, 2011):

$$\text{The improvement rate of CR } \left(IR_i\right) = \frac{\sum_{j=1}^{j}\left(IR_j * a_{i,j}\right)}{\left\{\sum_{j=1}^{j}\left(a_{i,j}\right)\right\} * W_i} \tag{2.2}$$

where

IR_i is the improvement rate of the customer requirement CR_i.

IR_j is the improvement rate of engineering metrics.

$a_{i,j}$ is the relation between customer voice i and engineering metric j.

W_i is the customer weight.

The improvement effect of the customer requirement is calculated by multiplying the improvement rate of the customer requirement (IR_i) by the customer weight W_i:

$$IE_i = IR_i * W_i$$

Then finally, the global score is calculated as the sum of the improvement effect of all customer requirements. The global score for all options are found and the one with the highest score implies good potential from a sustainable manufacturing viewpoint.

The scheme of ECQFD with four phases is depicted in Figure 2.1.

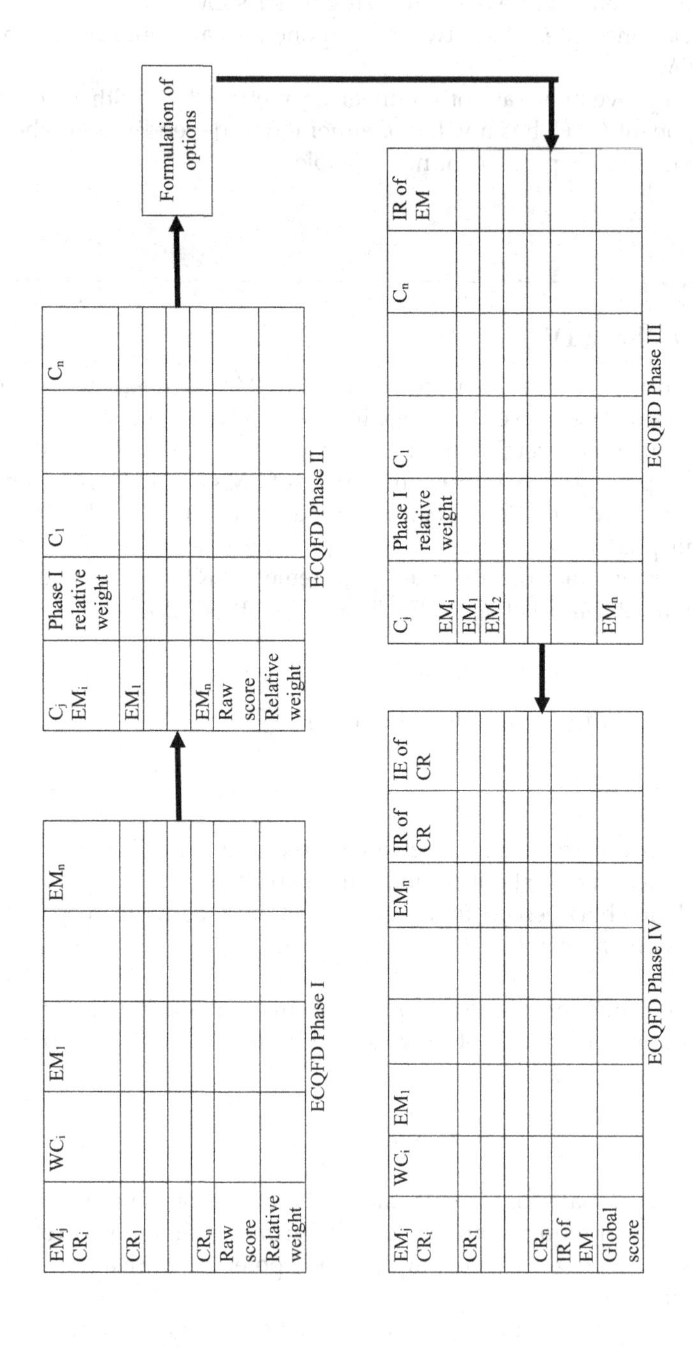

FIGURE 2.1
Scheme of ECQFD.

2.5 Extended Producer Responsibility (EPR) Policy

- The idea of extended producer responsibility (EPR) was integrated into the environmental policy by many governments in the 1990s from the perspective of increasing environmental challenges stipulated by unsustainable production and consumption by human beings (Yu et al., 2008).
- Initially announced in Germany pertaining to the Packaging Ordinance of 1991, the approach of EPR consequently spread around the world with regard to waste management (Fishbein, 1996).
- In 2001, then Organisation for Economic Co-operation and Development (OECD) defined EPR as 'an environmental policy approach where the producers' responsibility, physical and/or financial, for a product is prolonged to the post-consumer phase of a product's life cycle' (Xiang & Ming, 2011).
- The goal of EPR is to avoid environmental concerns at the source by facilitating incentives for modifications during the product design phase (Yu et al., 2008).

2.5.1 Unique Features of EPR Policy

1. The transfer of responsibility (physical and/or economical, full or partial) upstream to the manufacturer and away from municipalities.
2. Enabling incentives for manufacturers to inculcate environmental concerns in product design (OECD, 2001).

2.5.2 Potential Benefits of EPR Policy

- Sustainable raw material consumption.
- Sustainable product designs with longer duration.
- Eco-friendlier products (less hazardous and more easily recyclable) (Agamuthu & Victor, 2011).

2.6 Summary

This chapter details the procedure involved in the sustainable manufacturing tool environmentally conscious quality function deployment (ECQFD) in four phases which facilitate effective handling of traditional and environmental aspects in a scientific manner. Phase I is concerned with the identification of the priority of engineering metrics; phase II with the priority of components; phase III with the goal to compute the improvement rate of engineering metrics; and phase IV to compute the improvement effect of customer requirements. Also, the scope, features and benefits of an extended producer responsibility (EPR) policy are explained.

References

Agamuthu, P., & Victor, D. (2011). Policy trends of extended producer responsibility in Malaysia. *Waste Management and Research, 29*(9), 945–953.

Fishbein, B. (1996). Extended producer responsibility: A new concept spreads around the world. *Rutgers University Demanufacturing Partnership Program Newsletter, 1*(2). Retrieved from http://www.grrn.org/resources/Fishbein.html.

Masui, K., Sakao, T., Kobayashi, M., & Inaba, A. (2003). Applying Quality Function Deployment to environmentally conscious design. *International Journal of Quality and Reliability Management, 20*(1), 90–106.

Organisation for Economic Co-operation and Development (OECD). (2001). *Extended producer responsibility: A guidance manual for governments.* Paris: OECD Publishing.

Vinodh, S., & Rathod, G. (2011). Application of ECQFD for enabling environmentally conscious design and sustainable development in an electric vehicle. *Clean Technologies and Environmental Policy, 13*(2), 381–396.

Xiang, W., & Ming, C. (2011). Implementing extended producer responsibility: Vehicle remanufacturing in China. *Journal of Cleaner Production, 19*(6–7), 680–686.

Yu, J., Hills, P., & Welford, R. (2008). Extended producer responsibility and eco-design changes: Perspectives from China. *Corporate Social Responsibility and Environmental Management, 15*(2), 111–124.

3

Sustainable Manufacturing Tools: Life Cycle Assessment (LCA)

The life cycle assessment (LCA) is one of the potential tools of sustainable manufacturing. It enables the determination of environmental impacts across life cycle phases.

3.1 Fundamentals of LCA

LCA is an approach used to assess and evaluate the environmental impacts related to the product across its entire life (Ma et al., 2018). LCA can be done by referring to ISO 14040 (Principles and Framework) and ISO 14044 (Requirements and Guidelines) standards (Goedkoop et al., 2016).

3.1.1 Applications of LCA

LCA is applied for several goals (Goedkoop et al., 2016). Based on ISO standards for LCA, it helps in

- Identification of improvement scope for minimizing environmental impacts at several product life cycle phases.
- Enabling the decision maker to effectively make decisions in both government and non-government firms.

3.1.2 Steps of LCA

The four phases of LCA (Blengini, 2008) are shown in Figure 3.1.

- Goal and scope definition
- Life cycle inventory analysis
- Impact assessment
- Interpretation

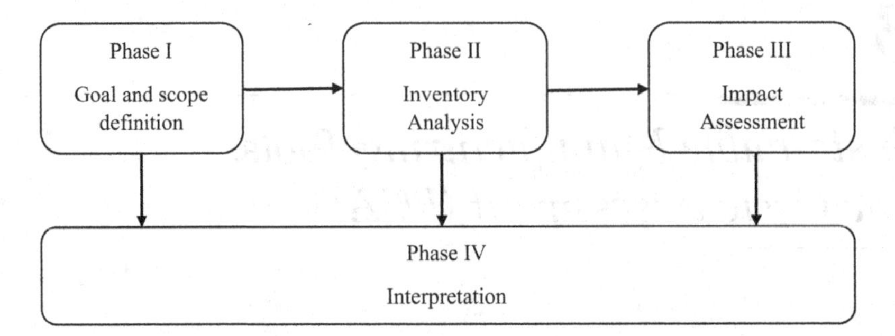

FIGURE 3.1
Life cycle assessment phases.

3.2 LCA Phase I: Goal and Scope Definition

In this phase, the objective of the study is mentioned along with its boundary conditions. The goal statement needs to be written and the scope defined (Goedkoop et al., 2016):

- The intended application of doing LCA, as it can be done for product development, product improvement, environmental declaration, etc.
- Description and definition of the product
- Assumptions and limitations of the study
- System boundaries of the study

3.2.1 Functional Unit

It is vital to specify the functional unit of the product. The functional unit will help in benchmarking of different products with similar functional units. It is defined to ensure that two products with different functional units are not compared. For example, the functional unit of a truck and computer cannot be compared.

3.2.2 System Boundaries

It is vital to specify the system boundaries of the study whether cradle to grave, cradle to gate, gate to grave or gate to gate (Böckin & Tillman, 2019). In this stage, the desirable phases need to be defined. It has to be specified whether LCA is to be done for the following boundary conditions.

There are four options for system boundaries (Blengini, 2008):

- *Cradle to grave:* This boundary option encompasses material extraction, processing, transportation, use and end-of-life phases.

- *Cradle to gate:* This boundary option includes material extraction and processing phases.
- *Gate to grave:* This boundary option includes production, transportation, use and end-of-life stages.
- *Gate to gate:* This boundary option includes the production phase only for LCA.

3.3 LCA Phase II: Inventory Analysis

This stage of LCA includes collecting data for performing analysis. The life cycle inventory stage involves collecting and quantifying all the input and output data for a product in its complete life cycle.

3.3.1 Types of Data Required in LCA

- *Foreground data:* These are the data that need to be collected and are specific for the product. It includes materials data and production data.
- *Background data:* These are the general standard data which are available in the software database and include data related to production of materials, energy, transportation, disposal actions etc. (Goedkoop et al., 2016).

3.4 LCA Phase III: Impact Assessment

Presently there are several evaluation approaches available for LCA. The selection of a specific method for assessment depends mainly on user requirements and geographical location.

The impact assessment phase assesses the quantum of environmental impacts generated during the life cycle phases being considered (Hauschild & Huijbregts, 2015). This phase consists of five elements: classification, characterization, normalization, grouping and weighting.

3.4.1 Classification

The results from the inventory analysis phase include various emissions. All these emissions are classified in different categories. In this step, first the different categories need to be identified based on user requirements, then all emissions are divided into the considered categories (Goedkoop et al., 2016). In some cases, certain emissions may influence more than one impact category. For example, NoX emissions contribute to both acidification and eutrophication.

3.4.2 Characterization

For every elementary flow for an impact category, the quantity is multiplied with the characterization factor (Hauschild & Huijbregts, 2015). The characterization factor provides the quantitative indication of its significance for a specific impact category.

3.4.3 Normalization

In normalization, the magnitude of impact needs to be assessed based on a particular reference (Seppälä & Hämäläinen, 2001). This can be done by comparing the impact category with the reference. In this phase, different impact categories are compared with a particular reference so that it could be written in a common scale of value.

3.4.4 Grouping

All the impact categories are sorted and ranked. This can be done in two different ways. The first way is to do sorting according to input and output data, based on regional factor. The second way of ranking is based on hierarchy (based on priority: high, medium and low). Ranking is based on the user; different industries/organizations may have different preferences based on which impact categories were ranked (Goedkoop et al., 2016).

3.4.5 Weighting

Weighting is an optional step in LCA and is based on the user. There is no scientific principle for weighting. It is used to compare different indicators based on their importance (Seppälä & Hämäläinen, 2001). It is a difficult step in LCA for midpoint methods. Certain solutions for weighting are indicated in Goedkoop et al. (2016).

3.5 LCA Phase IV: Interpretation

The interpretation stage is the last step of LCA and includes several tests to ensure the adequacy of the conclusions. In this phase, the results are analyzed to check the consistency with the goal and scope definition. It helps in ensuring completeness of the study (Goedkoop et al., 2016).

The results of LCA can be presented in bar graphs, lists and tables. They can be presented in a structured way based on different processes, impact categories and life cycle phases.

3.5.1 Evaluation

The purpose of evaluation is to check and improve the reliability of the study. Three methods used for evaluation are completeness, sensitivity and consistency checks. The details can be found in Goedkoop et al. (2016).

3.5.2 Conclusions, Recommendations and Reporting

The main objective of the interpretation phase of LCA is to derive conclusions from the study, to report limitations of the study and to contribute endorsements based on the evaluated results. The conclusions can

- Determine the vital issues.
- Assess the approach and results for completeness, sensitivity and consistency.
- Derive preliminary inferences in line with the study objectives.
- In line with consistency, final conclusions can be drawn.

3.5.3 Report

The results of an LCA study need to be offered in a comprehensive report in a structured manner (Goedkoop et al., 2016). The report should include inventory results and impact assessment results, along with the method considered and assumptions made in the study.

3.6 Life Cycle Cost Analysis

The life cycle costing procedure (Kaebernick et al., 2003; Rathod et al., 2011) is shown in Figure 3.2. This includes the comparison of forward and reverse manufacturing scenarios. In the forward manufacturing scenario, product cost is known, and environmental cost is found as 10% of the product cost. The product life cycle cost (PLCC) is the sum of product and environmental costs. The market price (MP) for components is known. The product effectiveness (PE) is a binary value (1, if component has feasibility for reverse

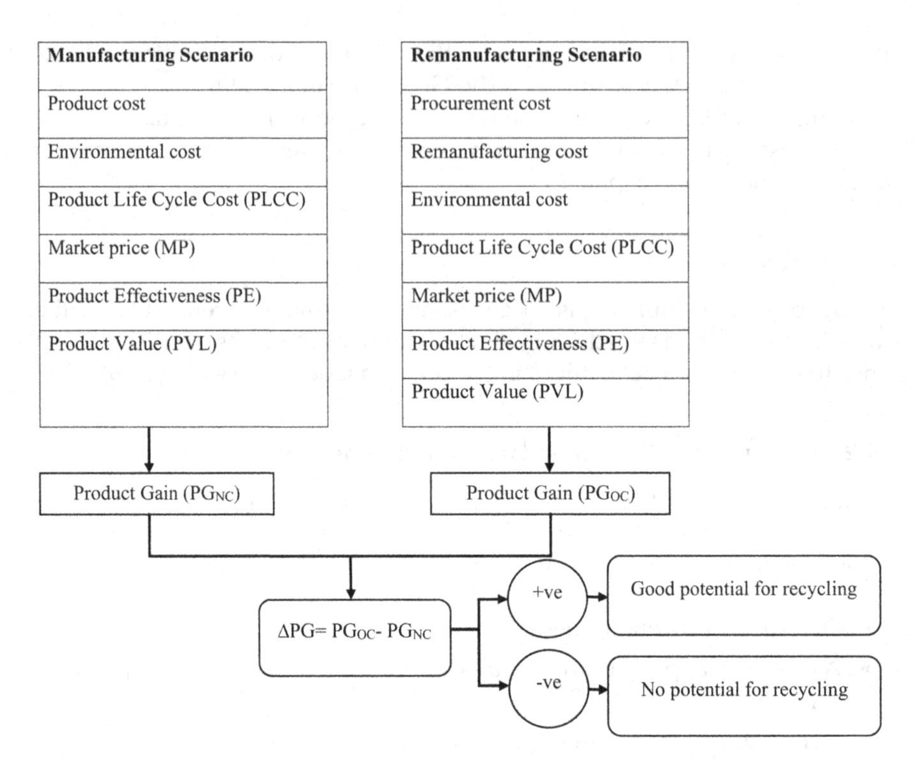

FIGURE 3.2
Life cycle costing procedure.

processing; 0, otherwise). The product value (PVL) is the product of MP and PE. PG_{NC} is the difference between PVL and PLCC. In the reverse manufacturing scenario, the procurement cost implies the cost associated with purchase of components from the customer, remanufacturing cost is known and the environmental cost is set to 1% of the remanufacturing cost (Kaebernick et al., 2003). PLCC is the sum of the three cost elements. MP, PE and PVL computations remain the same as in the forward manufacturing case. PG_{OC} is the difference between PVL and PLCC. If $\Delta PG = PG_{OC} - PG_{NC}$ is positive, the component has good potential for reverse processing; otherwise the potential is less. Equations used for computation of LCC are presented next.

Product gain for manufacturing a new component can be calculated as

$$\text{Product Life Cycle Cost (PLCC)} = \text{Product cost} + \text{Environmental cost} \tag{3.1}$$

$$\text{Product Value (PVL)} = \text{Market price (MP)} * \text{Product Effectiveness (PE)} \tag{3.2}$$

$$\text{Product Gain } (PG_{NC}) = \text{Product Value } (PVL) - \text{Product Life Cycle Cost } (PLCC) \tag{3.3}$$

Product gain for remanufacturing an old part can be calculated as

$$\text{Product Life Cycle Cost (PLCC)} = \text{Procurement cost} + \text{Remanufacturing cost} + \text{Environmental cost} \tag{3.4}$$

$$\text{Product Value (PVL)} = \text{Market price (MP)} \times \text{Product Effectiveness (PE)} \tag{3.5}$$

$$\text{Product Gain } (PG_{OC}) = \text{Product Value (PVL)} - \text{Product Life Cycle Cost (PLCC)} \tag{3.6}$$

Hence, net product gain is

$$\Delta PG = PG_{OC} - PG_{NC} \tag{3.7}$$

3.7 Summary

This chapter provides readers with the fundamentals and application steps of a life cycle assessment (LCA) in four phases: goal and scope definition, inventory analysis, impact assessment, and interpretation. The details of the four phases are discussed. The procedure of life cycle costing with the computations are also presented. The scenarios of manufacturing a new component and reverse manufacturing an old component are discussed.

References

Blengini, G.A. (2008). Using LCA to evaluate impacts and resources conservation potential of composting: A case study of the Asti District in Italy. *Resources, Conservation and Recycling*, *52*(12), 1373–1381.

Böckin, D., & Tillman, A.M. (2019). Environmental assessment of additive manufacturing in the automotive industry. *Journal of Cleaner Production*, *226*, 977–987.

Goedkoop, M., Oele, M., Leijting, J., Ponsioen, T., & Meijer, E. (2016). Introduction to LCA with SimaPro. PRé.

Hauschild, M.Z., & Huijbregts, M.A. (2015). Introducing life cycle impact assessment. In Hauschild, Michael Z., and Huijbregts, Mark A.J. (eds.), *Life cycle impact assessment* (pp. 1–16). Dordrecht: Springer.

Kaebernick, H., Kara, S., & Sun, M. (2003). Sustainable product development and manufacturing by considering environmental requirements. *Robotics and Computer-Integrated Manufacturing, 19*(6), 461–468.

Ma, J., Harstvedt, J.D., Dunaway, D., Bian, L., & Jaradat, R. (2018). An exploratory investigation of additively manufactured product life cycle sustainability assessment. *Journal of Cleaner Production, 192*, 55–70.

Rathod, G., Vinodh, S., & Madhyasta, U.R. (2011). Integration of ECQFD and LCA for enabling sustainable product design in an electric vehicle manufacturing organisation. *International Journal of Sustainable Engineering*, 4(3), 202–214.

Seppälä, J., & Hämäläinen, R.P. (2001). On the meaning of the distance-to-target weighting method and normalisation in life cycle impact assessment. *The International Journal of Life Cycle Assessment*, 6(4), 211.

4

Life Cycle Impact Assessment Methods: CML, Eco-Indicator 95 and Eco-Indicator 99

4.1 CML Method

The details of the CML method are presented in the following sections.

4.1.1 Description

CML 2001 is a life cycle impact assessment (LCIA) approach which is based on quantitative modelling so as to minimize uncertainty at the early stage in the cause–effect chain. The resultant impact categories are grouped in midpoint categories based on common mechanisms (Guinée et al., 2002).

The CML method for damage assessment was proposed by the University of Leiden in the Netherlands in 2001 considering more than 1700 different flows (Acero et al., 2014).

The CML method is grouped in two different categories, namely baseline and non-baseline. The baseline impact category includes impact categories which are most common in LCA studies, whereas non-baseline categories are operational impact categories which depend on the requirement (Menoufi, 2011).

Some of the impact categories in the baseline type are acidification potential, climate change, eutrophication potential and human toxicity (Hischier et al., 2010).

The non-baseline or optional impact categories are those which are required based on specific studies. Impact categories under this type can be found in Hischier et al. (2010).

Characterization in the CML method is in line with European average values.

Normalization in the CML method is based on the global normalization factor pertaining to 1990 and 1995 as aggregate annual world interventions. The weighting method is not considered in the CML method.

4.1.2 Normalization

Normalization factors for CML 2001 exist for the European Union (EU) and the world.

The normalization factors for the CML method are assessed in line with total substance emissions and the characterization factor per substance. Normalization data for different geographical areas can be scaled based on the gross domestic product of original CML 2001 normalization.

The normalization factor (N.F) of any impact category can be evaluated by the product of its characterization factors (C.F) and their corresponding emissions. For every impact category, the sum of the product value provides the normalization factor.

$$N.F = C.F \times \text{Substance emissions} \qquad \text{(Hischier et al., 2010)} \qquad (4.1)$$

To yield the normalized result from the characterized results, the characterization factor is divided by the normalization factor.

4.1.3 Weighting

Weighting is a non-compulsory step in CML. No baseline method was proposed.

4.1.4 Features of CML

- It includes a characterization factor for all characterization methods of LCA.
- It contains normalization data for all impact categories and interventions at different levels.
- There is a difference between baseline and non-baseline (study-specific) impact groups (JRC-IES, 2010).

The time horizon is infinite for the CML method. But in some cases, it is suggested to perform the analysis for a shorter time horizon so as to minimize uncertainty associated with the toxic effect of materials (JRC-IES, 2010).

The CML is based on midpoint impact assessment; however, the relation between the midpoint and endpoint is specified.

Approximately 800 substances are covered in the CML method (JRC-IES, 2010).

The details of the characterization factors pertaining to emissions in air, water, soil, resource use and land use can be found in Hischier et al. (2010).

4.1.5 Quality Considerations

Forty-eight percent of the basic flows in the ecoinvent database have an associated basic flow in CML 2001. But for the majority of these flows, no CFs were indicated. Thus only for 20% of the basic flows, a characterization factor other than zero is deployed (Hischier et al., 2010).

4.2 Eco-Indicator Method

The Eco-indicator method is a weighting approach for evaluating environmental impact considering damages to resources, human health and the ecosystem. It consists of 100 figures for 100 different processes (Goedkoop et al., 1995).

4.2.1 Principles of the Eco-Indicator

The Eco-indicator method is a vital tool that helps the designer to analyze the most environment-friendly design by considering various aspects (Goedkoop et al., 1995).

4.2.2 Scope of Eco-Indicator

1. The LCA method is upgraded by considering the weighting method. This method helps in calculating one single score for environmental impact by considering all possible effects.
2. Data exists for most used materials and processes. The Eco-indicator score needs to be computed from this.

In LCA, the Eco-indicator is a number which indicates the environmental impact associated with the process or material. The higher the value of the indicator, the more the environment will be impacted. The Eco-indicator enables environmental impact evaluation within the scope of the designer.

4.2.3 Environmental Effects

Environmental effects considered include the greenhouse effect, ozone layer depletion, acidification, eutrophication, smog and toxic substances.

The details of environmental effects not included can be found in Goedkoop et al. (1995).

4.2.4 Operating Instructions

While using the Eco-indicator method, the necessary steps to be followed for ensuring the correctness of application are provided next (Goedkoop et al., 1995).

Step 1: Institute the goal of the Eco-indicator computation.

- Describe the product or part which is considered.
- Define what is to be done. Perform an analysis of the product or a benchmark of the product with another product.
- Mention the accuracy level needed.

Step 2: State the life cycle.

- Prepare a product life cycle by considering all aspects including production, use and disposal.

Step 3: Quantify materials and processes.

- Analyze a functional unit.
- Quantify all processes.
- Assume data for missing observations.

Step 4: Input the details.

- Specify the materials and process details along with their amounts.
- Identify the necessary Eco-indicator values and specify them.
- Compute the scores by the product of the content and the indicator values.
- Cumulate the secondary results together.

Step 5: Infer the results.

- Integrate inferences with the results.
- Verify the impact of assumptions and uncertainties.
- Improve learnings (if necessary).
- Verify whether the objective of the computation is complied with.

4.2.5 Description of the Eco-Indicator 95 Method

Eco-indicator 95 follows the impact-oriented approach and is based on distance to target. It is a midpoint-based method. It is in line with the weighting and reduction factors of Dutch environmental policy (Goedkoop et al., 1995). All relevant processes were analyzed and all emissions were consolidated in an impact table. Environmental impacts are calculated based on the impact table. It is reasonably comprehensive and enables a consistent process for weighting.

4.3 Eco-Indicator 99 Method

The Eco-indicator 99 method is a widely used LCIA method. It is an upgrade of Eco-indicator 95, the initial endpoint evaluation approach. Eco-indicator 99 helps in assessing the environmental impact in a single score (Acero et al., 2014).

This method analyzes the environmental impact in three damage categories: human health, resources and ecosystem quality. The important point to be noted in this method is it provides a common unit, i.e. a standard unit for all categories. It is represented as point (Pt) or millipoint (mPt). As this method is mainly used for comparison of products or components, the value generated by this method is more relevant for comparison (Acero et al., 2014).

4.3.1 Different Archetypes in Eco-Indicator 99

- H – Hierarchist (default)
- I – Individualist
- E – Egalitarian

Comparison of the three archetypes of the Eco-indicator based on weightage to the damage categories (Hischier et al., 2010) is depicted in Table 4.1.

The method encompasses various normalization and weighting factors for several perspectives. The characterization factors for Eco-indicator 99 were available in CML 2001 documentation (Acero et al., 2014).

4.3.2 Features of Eco-Indicator 99

- Extension of Eco-indicator 95.
- Purpose – endpoint method with a focus on panel weighting.
- Midpoint impacts can be found in JRC-IES (2010).
- Endpoint impacts – human health, ecosystem quality, resources depletion.
- Substances covered – approximately 391.
- Weighting method – panel, weighting triangle or monetization method.
- Damage to human health, ecosystem quality and resources can be found in JRC-IES (2010).

TABLE 4.1

Benchmark of Three Archetypes of the Eco-Indicator Method

	Damage Category		
Archetype	**Human Health**	**Ecosystem**	**Resources**
Hierarchist	40%	40%	20%
Individualist	55%	25%	20%
Egalitarian	30%	50%	20%

4.4 ReCiPe Method

The details of the ReCiPe method are presented in the following sections.

4.4.1 Overview

ReCiPe is an LCIA method which provides results both at a midpoint level and endpoint level. The name ReCiPe is derived from its main contributors, namely RIVM and Radboud University, CML, and PRé (Goedkoop et al., 2009).

The ReCiPe method is a combined approach which includes both CML 2002 and Eco-indicator 99. It integrates the midpoint impact approach and endpoint indicators in a consistent framework. Initially it integrates both methods and then all impact categories are redeveloped and updated (Acero et al., 2014).

ReCiPe 2008 includes two sets of impact groups with related sets of characterization factors. Eighteen impact categories are dealt with at the midpoint level. At the endpoint level, these midpoint impact groups are further transformed and combined into three endpoint groups: damage to human health, ecosystem diversity and resource availability. The indicators pertaining to impact categories can be found in Goedkoop et al. (2009).

In ReCiPe, both the midpoint and endpoint methods can be implemented. In the midpoint method, only three damage categories will be considered, namely human health, ecosystem and resources, and by considering different perspectives, i.e. midpoint level individualist (I), egalitarian (E) and hierarchist (H) (Acero et al., 2014).

In the endpoint method, various indicators of the midpoint are also considered along with three damage categories. By implementing normalization and weighting methods, a single score is also calculated in this method. In normalization and weighting methods, normalization values from Europe and average weighting factors are considered (Goedkoop et al., 2009).

4.4.2 Connectivity between Midpoint and Endpoint Levels

The purpose of ReCiPe 2008 is to integrate two LCA methods, namely midpoint-based CML 2001 and endpoint-based Eco-indicator 99 (Dong & Ng, 2014).

ReCiPe 2008 includes all three types of characterization factors.

4.4.3 Dealing with Uncertainties and Assumptions: Scenarios

Similar to the Eco-indicator 99 method, in ReCiPe 2008 various sources of uncertainty are also grouped into limited scenarios (individualist, hierarchist, egalitarian).

4.4.4 Unique Features of ReCiPe 2008

- Reliable use of midpoints and endpoints in the common environmental mechanism.
- Reliable marginal method.
- Most impact types are elaborated in peer-reviewed papers (JRC-IES, 2010).

The characterization factors for midpoint and endpoint impacts are calculated based on the environmental cause–effect chain.

4.4.5 Normalization in ReCiPe 2008

Normalization data are available for European regions for 16 midpoint impact categories along with three endpoint impact categories (JRC-IES, 2010).

4.4.6 Weighting in ReCiPe 2008

Weighting methods are provided in JRC-IES (2010).

- For endpoints, a manual for panel weights exists.
- For midpoints, a monetization factor in line with prevention costs is indicated.
- For endpoints, a monetization factor in line with damage costs is indicated.
- At the endpoint level, the weighting triangle can be applied.

4.5 Summary

The chapter provides details regarding life cycle impact assessment (LCIA) methods: CML, Eco-Indicator 95, Eco-Indicator 99 and ReCiPe. The description of the methods covers details including normalization, weighting, impacts and time horizon.

References

Acero, A.P., Rodríguez, C., & Ciroth, A. (2014). *LCIA methods–Impact assessment methods in life cycle assessment and their impact categories*. GreenDelta GmbH, Berlin, Germany, p. 23.

Dong, Y.H., & Ng, S.T. (2014). Comparing the midpoint and endpoint approaches based on ReCiPe—A study of commercial buildings in Hong Kong. *The International Journal of Life Cycle Assessment, 19*(7), 1409–1423.

Goedkoop, M., Demmers, M., & Collignon, M. (1995). *The Eco-indicator 95: Manual for designers.* Pré Consultants, The Netherlands.

Goedkoop, M., Heijungs, R., Huijbregts, M., De Schryver, A., Struijs, J., & Van Zelm, R. (2009). ReCiPe 2008. *A life cycle impact assessment method which comprises harmonised category indicators at the midpoint and the endpoint level, 1,* 1–126.

Guinée, J.B., Gorrée, M., Heijungs, R., Huppes, G., Kleijn, R., Koning, A. de, … Huijbregts, M.A.J. (2002). *Handbook on life cycle assessment. Operational guide to the ISO standards. I: LCA in perspective,* 692 pp. Dordrecht: Kluwer Academic Publishers. ISBN 1-4020-0228-9.

Hischier, R., Weidema, B., Althaus, H.J., Bauer, C., Doka, G., Dones, R., ... & Köllner, T. (2010). *Implementation of life cycle impact assessment methods. Ecoinvent report no. 3, v2. 2.* Dübendorf, Switzerland: Swiss Centre for Life Cycle Inventories. Retrieved from https://www.universiteitleiden.nl/en/research/research-projects/science/cml-new-dutch-lca-guide.

European Commission, Joint Research Centre, Institute for Environment and Sustainability (JRC-IES) (2010). *ILCD handbook: Analysis of existing environmental impact assessment methodologies for use in life cycle assessment,* Joint Research Centre European Commission, Italy.

Menoufi, K.A.I. (2011). *Life cycle analysis and life cycle impact assessment methodologies: A state of the art,* Masters Degree Thesis, University in Lleida, Spain.

5

Design Strategies Supporting Sustainable Manufacturing

This chapter deals with four design strategies enabling sustainable manufacturing: design for disassembly, design for recycling, design for environment and eco-friendly product design.

5.1 Design for Disassembly

Disassembly is an effective approach to segregate a product into its essential parts, components, subassemblies or other groupings for various reasons (Güngör, 2006). Design for disassembly (DFD) gains importance in the context of sustainable manufacturing as it is a vital process facilitating end-of-life (EoL) management of products.

5.1.1 Principles

One of the principles of DFD is to use as few parts as possible in the product, which enables reduction of the time required to access desired part(s) of the product and to reassemble the product. In product recovery, it minimizes disassembly time and enables selective separation of materials. Also, the number of fasteners and tools needed for disassembly/reassembly tasks are significantly minimized (Güngör, 2006).

5.1.2 Scope

Disassembly tasks in both product recovery and maintenance are impacted during the product design stage. The number of parts and fasteners used to design the product greatly affects the efficiency of the disassembly process (Güngör, 2006). Thus, DFD is important and hence is given intense focus.

5.1.3 Need

There are three vital reasons for disassembly (Sodhi et al., 2004), namely product service or repair, end-of-life segregation of parts and elimination of toxic parts.

5.1.4 Benefits

DFD efforts lead to recognizing design attributes to reduce the product structure complexity by minimizing the number of parts, enhancing the usage of common materials, and selecting fasteners and joints which can be separated with ease (Güngör, 2006).

5.1.5 Factors Affecting Disassembly

The following factors affect disassembly (Desai & Mital, 2005):

- Use of less force
- A simple mechanism
- Fewer tools/simple processes
- Minimization of parts repetition
- Identification of disassembly points
- Simple product structure

5.1.6 Types of Disassembly

The disassembly process is shown in Figure 5.1. It has three stages: pre-process, in-process and after process. Disassembly can be divided into disassembly with or without force. Disassembly directions are also vital to understanding the disassembly mechanism (Mok et al., 1997).

Attributes of the disassembly process can be recognized by dividing the disassembly process based on disassembly time and sequence (Mok et al., 1997).

The types of disassembly are shown in Figure 5.2. Disassembly is done either by destructive or non-destructive action in line with the objective.

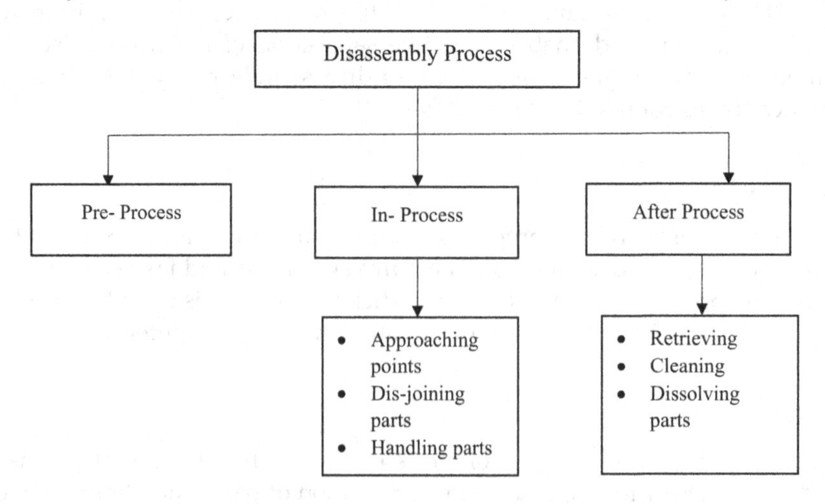

FIGURE 5.1
Disassembly process.

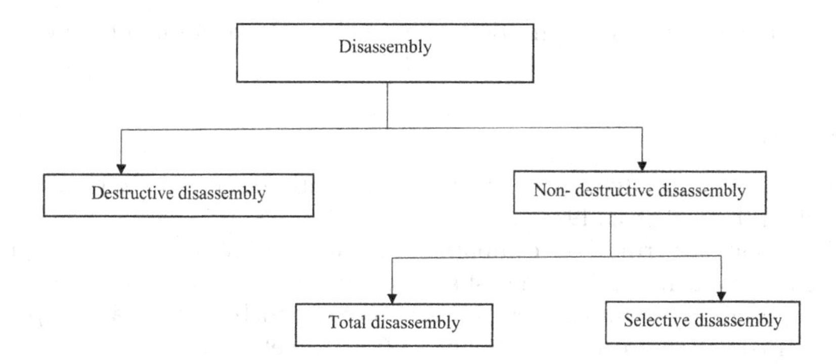

FIGURE 5.2
Types of disassembly.

With reference to the degree of disassembly, non-destructive disassembly is categorized into two types (Desai & Mital, 2003):

- Total disassembly
- Selective disassembly

5.1.7 Designing for Active Disassembly

Active disassembly is a process wherein a product utilizes an external trigger, e.g. temperature, for the release of fasteners. It involves utilizing smart materials, and parts undergo self-disassembly when exposed to external triggering factors (Soh et al., 2014).

Successful DFD includes three major aspects (Bogue, 2007):

- Collection and utilization of materials
- Component design and product architecture
- Collection and utilization of joints, connectors and fasteners

5.2 Design for Recycling

Design for recycling (DfR) is designing a recyclable product and utilizing recycled materials to replace virgin materials (Maris et al., 2014). DfR has been a key focus of environmental management for computer manufacturers in recent years. "Takeback" policies in Europe and other initiatives focused the computer industry on continuous recyclability enhancement of its product designs. A recyclable material must sustain its mechanical and chemical properties, and facilitate sorting by recycling firms. Eco designing recyclable materials implies making them convertible and separable with the recommended cost-to-performance ratio (Maris et al., 2014).

DfR includes materials applications that improve recyclability of the whole product or a part of it (Ardente et al., 2003).

5.2.1 Scope

The most ecologically effective way to cope with an end-of-life product is recycling (Kriwet et al., 1995).

Two issues are critical for evaluation. First is the need to integrate several design goals such as minimal cost and reduced waste. Second is the uncertainty pertaining to future recycling stipulations, such as raw material price and refinement of process technologies (Zussman et al., 1994).

5.2.2 Need

Recycling and remanufacturing offer options to waste generation and material consumption. Recycling of parts and materials minimizes the requirement for virgin material, thus minimizing the waste related to extraction (Hundal, 2000).

5.2.3 Recycling Process

Recycling focusses on 'closing the loop' of materials and components after utilization by (re)using/using them for new products. Three loops can be differentiated in a product life cycle in which recycling tasks can be done (Kriwet et al., 1995): recycling of production scraps, during product usage and after product usage.

5.2.4 Types of Recycling

Table 5.1 depicts types of recycling. The European Union provides best available techniques (BAT) reference documents that include important information about several technologies (Reuter, 2011).

5.2.5 Benefits

- Eco-friendlier products
- Cost minimization and waste reduction

5.3 Design for Environment (DFE)

Design for environment (DFE) is an effective amalgamation of environmental concerns into industrial product and process design (Young & Rollefson, 2000). Eco design or DFE addresses any design activity which focuses on enhancing the product's environmental performance (Hauschild

TABLE 5.1

Types of Recycling

Research study	Type of Recycling
Cui and Forssberg (2003), Froelich et al. (2007), Siddique et al. (2008)	Physical recycling • Mechanical e-waste recycling • Plastics sorting and recycling • Recycling plastics into concrete
Helsen et al. (1998)	Wood recycling
Xua et al. (2008), Muller and Friedrich (2006), Bernardes et al. (2004), Briffaerts et al. (2009)	Battery recycling • Lithium battery recycling • Nickel-metal hydride battery recycling • General battery recycling • Specific battery recycling processes
Shent et al. (1999), Guo et al. (2009)	Plastics recycling • Recycling of non-metallic portions of printed wire boards

et al., 2004). DFE implies product designs that reduce environmental impact across their life cycle, including raw material extraction till end of life (Keoleian & Menerey, 1994).

5.3.1 Need for DFE

There is a continuous need to explore the impact of design across the complete product life cycle. The importance is based on the systematic character of the proposed method (Belvedere & Grando, 2017).

5.3.2 Practices

There are several DFE practices related with eco-efficient design. Ideally, a single design innovation may relate to fulfilling various types of goals. For example, reduction in product mass can result in minimal energy and material utilization, and pollutant emission reduction.

Some of the common DFE practices are

- Material substitution
- Waste source/reduced energy use
- Life extension
- Design for separability and disassembly, recyclability, and remanufacture

5.3.3 Benefits

- Reduced resource consumption and production costs
- Reduced resource utilization and waste, and improved product reusability

- Facilitate remanufacturing
- Reduced remanufacturing cost (Zheng et al., 2019)

5.3.4 Eco Design Strategy Wheel

The steps used in the approach are as follows (Young & Rollefson, 2000):

1. Select low-environmental-impact materials.

   This step focuses on selecting the most eco-friendly relevant material for the application such as material durability or recycling potential.
2. Reduce material quantity.

   The necessity for less material derives from lean but strong designs by utilizing lightweight materials.
3. Optimize production techniques.

   Refine production or explore alternative approaches to enhance energy efficiency, and reduce waste and pollution.
4. Optimize distribution system.

   Consider packaging needs of a product and efficiency of transport to its eventual market.
5. Reduce impact during use.

   Reduce the need for consumables – such as energy, water and detergent – and limit the need for auxiliary materials, like batteries.
6. Optimize initial lifetime.

   Explore the maximum use of a product by ensuring durability and ease of repair.
7. Optimize end-of-life system.

   Adapt the product for ease of disassembly.
8. Concept development.

   Rethink the essential function(s) to be contributed by a product or service, and assess whether innovative approaches may contribute to vital enhancement.

5.3.5 DFE Process

The design for environment implementation process across various phases is indicated in the following (Moreira et al., 2018; Belvedere & Grando, 2017):

- Product planning
- Base design

- Embodiment design
- Detail design
- Test and validation
- Manufacturing preparation

Design guidelines pertaining to sustainability can be applied from the base design stage to realize the product concept.

5.4 Eco-Friendly Product Design Methods

Eco design is a product design and development method that aims to attain a reasonable balance among economic and environmental benefits (Veshagh & Li, 2006). Eco-friendly product development strategies enable product development effectiveness advantages for firms (Katsikeas et al., 2016).

5.4.1 Definition

Eco-friendly product design is commonly viewed as the amalgamation of environmental concerns into the traditional design process from cradle to grave (Veshagh & Li, 2006).

5.4.2 Scope

- Minimize the risk of pollutant emissions that violate climate change conventions (Kim et al., 2014).
- Facilitate companies to comply with environmental regulations, attain competitive advantage and contribute to future growth (Katsikeas et al., 2016).
- Improve the product environmental impact during various life cycle phases (Katsikeas et al., 2016).

5.4.3 Need

Eco efficiency has a greater role in the initial design stages, such as planning and concept design phases. Hence, it is critical that the environmental issues be considered in initial design phases (Kobayashi, 2006).

Environmental concerns are vital to product designers and manufacturers (Mok et al., 2006). Thus, environmentally friendlier design approaches need to be derived and amalgamated for a product's complete life cycle.

5.4.4 Tools That Include Environmental Needs in the Design Process

5.4.4.1 Tools Based on Design Matrix

Tools based in the design matrix include descriptive methods that are concerned with the qualitative assessment of the design team for various product needs across its life cycle (Bovea & Pérez-Belis, 2012).
Tools:

- Requirements matrix
- DFE matrix

5.4.4.2 Tools Based on Quality Function Deployment

Several tools allow quality function deployment (QFD) to be implemented in order to include environmental needs during the initial product design phases. These tools are used to verify that the product fulfils customer needs and environmental needs.

Some of the tools are (Bovea & Pérez-Belis, 2012)

- Green-QFD (GQFD)
- Environmental-QFD (E-QFD)
- House of ecology (HoE)
- QFD for environment (QFDE)

5.4.4.3 Tools Based on Value Analysis (VA)

Tools in line with value analysis (VA) enable the design/redesign of a product at a lesser cost, and consider customers' willingness to pay for apparent environmental advantages.
Tools:

- Life cycle environmental cost analysis (LCECA)
- Eco-value analysis (Eco-VA)
- Eco-redesign

5.4.4.4 Tools Based on Failure Mode and Effect Analysis (FMEA)

The failure mode and effect analysis (FMEA) approach is applied to recognize, evaluate and avoid concerns connected with product safety.
Tools:

- Environmental FMEA (E-FMEA)
- Eco-FMEA

5.4.4.5 Other Tools

Other approaches are based on the theory of inventive problem solving (TRIZ) and the Kano model.

5.4.5 Review of Eco Design Strategies and Methodologies

- Design for recyclability/remanufacturing
- Design to minimize material usage
- Design for disassembly/durability
- Design for energy efficiency (Veshagh & Li, 2006)

5.4.6 Steps Involved in Eco-Innovative Product Design

Eco innovative product design in line with life cycle planning includes the following steps (Kobayashi, 2006).

Step 1: Life cycle planning (LCP). It includes product family planning, target specification, preliminary estimation, idea generation and concept evaluation at the product and component levels to generate design concepts at the component level to complete the LCP process.

Step 2: Embodiment and detail design.

Next, design engineers arrive at the arrangement, final structure and product shape on a computer-aided design (CAD) platform in line with eco-design concept (step 2). Hence, the prevailing design for excellence (DfX) tool, such as design for assembly, disassembly and recyclability, can be utilized.

Step 3: LCA computation.

Step 4: Factor-X computation.

Ultimately, the efficiency of the eco-design is validated using LCA (step 3) and the factor-X indicator (step 4).

5.4.7 Benefits of Eco Design Strategies

- Enhanced environmental performance
- Preparedness for compliance with future regulations
- Lower manufacturing costs
- Lower product costs
- Enhanced market access/share
- Improved long-term sustainability Veshagh & Li, 2006)

5.5 Summary

The chapter provides readers with various design strategies supporting sustainable manufacturing. The scope, need, process and benefits of designing for disassembly are presented. The disassembly process, types and factors affecting disassembly are discussed. The need for recycling, recycling types and methods are presented. The need for DFE, steps, processes and benefits of DFE are discussed. Tools in various categories, benefits and steps of eco-friendly product design are discussed.

References

Ardente, F., Beccali, G., & Cellura, M. (2003). Eco-sustainable energy and environmental strategies in design for recycling: The software "ENDLESS". *Ecological Modelling, 163*(1–2), 101–118.

Belvedere, V., & Grando, A. (2017). *Sustainable operations and supply chain management.* Chichester: John Wiley & Sons.

Bernardes, A.M., Espinosa, D.C.R., & Tenório, J.S. (2004). Recycling of batteries: A review of current processes and technologies. *Journal of Power Sources, 130*(1–2), 291–298.

Bogue, R. (2007). Design for disassembly: A critical twenty-first century discipline. *Assembly Automation, 27*(4), 285–289.

Bovea, M., & Pérez-Belis, V. (2012). A taxonomy of ecodesign tools for integrating environmental requirements into the product design process. *Journal of Cleaner Production, 20*(1), 61–71.

Briffaerts, K., Spirinckx, C., Van der Linden, A., & Vrancken, K. (2009). Waste battery treatment options: Comparing their environmental performance. *Waste Management, 29*(8), 2321–2331.

Cui, J., & Forssberg, E. (2003). Mechanical recycling of waste electric and electronic equipment: A review. *Journal of Hazardous Materials, 99*(3), 243–263.

Desai, A., & Mital, A. (2003). Evaluation of disassemblability to enable design for disassembly in mass production. *International Journal of Industrial Ergonomics, 32*, 265–281.

Desai, A., & Mital, A. (2005). Incorporating work factors in design for disassembly in product design. *Journal of Manufacturing Technology Management, 16*(7), 712–732.

Froelich, D., Maris, E., Haoues, N., Chemineau, L., Renard, H., Abraham, F., & Lassartesses, R. (2007). State of the art of plastic sorting and recycling: Feedback to vehicle design. *Minerals Engineering, 20*(9), 902–912.

Güngör, A. (2006). Evaluation of connection types in design for disassembly (DFD) using analytic network process. *Computers and Industrial Engineering, 50*(1–2), 35–54.

Guo, J., Guo, J., & Xu, Z. (2009). Recycling of non-metallic fractions from waste printed circuit boards: A review. *Journal of Hazardous Materials, 168*(2–3), 567–590.

Hauschild, M.Z., Jeswiet, J., & Alting, L. (2004). Design for environment—Do we get the focus right? *CIRP Annals, 53*(1), 1–4.

Helsen, L., Van den Bulck, E., & Hery, J.S. (1998). Total recycling of CCA treated wood waste by low-temperature pyrolysis. *Waste Management, 18*(6–8), 571–578.

Hundal, M. (2000, May). Design for recycling and remanufacturing. In *International design conference*, Dubrovnik-Cavtat, Croatia.

Katsikeas, C.S., Leonidou, C.N., & Zeriti, A. (2016). Eco-friendly product development strategy: Antecedents, outcomes, and contingent effects. *Journal of the Academy of Marketing Science, 44*(6), 660–684.

Keoleian, G.A., & Menerey, D. (1994). Sustainable development by design: Review of life cycle design and related approaches. *Air and Waste, 44*(5), 645–668.

Kim, J.Y., Jeong, S.J., Cho, Y.J., & Kim, K.S. (2014). Eco-friendly manufacturing strategies for simultaneous consideration between productivity and environmental performances: A case study on a printed circuit board manufacturing. *Journal of Cleaner Production, 67*, 249–257.

Kobayashi, H. (2006). A systematic approach to eco-innovative product design based on life cycle planning. *Advanced Engineering Informatics, 20*(2), 113–125.

Kriwet, A., Zussman, E., & Seliger, G. (1995). Systematic integration of design-for-recycling into product design. *International Journal of Production Economics, 38*(1), 15–22.

Maris, E., Froelich, D., Aoussat, A., & Naffrechoux, E. (2014). From recycling to eco-design. In Ernst Worrell and Markus Reuter (Eds.), *Handbook of recycling* (pp. 421–427). Elsevier, The Netherlands.

Mok, H.S., Cho, J.R., & Moon, K.S. (2006, May). Design for environment-friendly product. In *International conference on computational science and its applications* (pp. 994–1003). Berlin, Heidelberg: Springer.

Mok, H.S., Kim, H.J., & Moon, K.S. (1997). Disassemblability of mechanical parts in automobile for recycling. *Computers and Industrial Engineering, 33*(3–4), 621–624.

Moreira, A.C., Ferreira, L.M.D., & Zimmermann, R.A. (Eds.). (2018). *Innovation and supply chain management: Relationship, collaboration and strategies*. Berlin: Springer.

Müller, T., & Friedrich, B. (2006). Development of a recycling process for nickel-metal hydride batteries. *Journal of Power Sources, 158*(2), 1498–1509.

Reuter, M.A. (2011). Limits of design for recycling and "sustainability": A review. *Waste and Biomass Valorization, 2*(2), 183–208.

Shent, H., Pugh, R.J., & Forssberg, E. (1999). A review of plastics waste recycling and the flotation of plastics. *Resources, Conservation and Recycling, 25*(2), 85–109.

Siddique, R., Khatib, J., & Kaur, I. (2008). Use of recycled plastic in concrete: A review. *Waste Management, 28*(10), 1835–1852.

Sodhi, R., Sonnenberg, M., & Das, S. (2004). Evaluating the unfastening effort in design for disassembly and serviceability. *Journal of Engineering Design, 15*(1), 69–90.

Soh, S.L., Ong, S.K., & Nee, A.Y.C. (2014). Design for disassembly for remanufacturing: Methodology and technology. *Procedia CIRP, 15*, 407–412.

Veshagh, A., & Li, W. (2006). Survey of eco design and manufacturing in automotive SMEs. In *Proceedings of the LCE 2006*, Leuven, Belgium (pp. 305–310).

Xua, J., Thomas, H.R., Francis, R.W., Lum, K.R., Wang, J., & Liang, B. (2008). A review of processes and technologies for the recycling of lithium-ion secondary batteries. *Journal of Power Sources, 177*(2), 512–527.

Young, S.B., & Rollefson, J. (2000). Design for environment. *Alternatives Journal, 26*(1), 36–37. Retrieved from https://search.proquest.com/docview/218760986?accountid=49696.

Zheng, X., Govindan, K., Deng, Q., & Feng, L. (2019). Effects of design for the environment on firms' production and remanufacturing strategies. *International Journal of Production Economics, 213,* 217–228.

Zussman, E., Kriwet, A., & Seliger, G. (1994). Disassembly-oriented assessment methodology to support design for recycling. *CIRP Annals, 43*(1), 9–14.

6

Standards for Sustainable Manufacturing

6.1 ISO 14001 Environmental Management Systems (EMS)

ISO 14001 is a global standard for environmental management systems (EMS) and includes management processes that necessitate organizations to recognize, quantify and govern their environmental impacts (Bansal & Hunter, 2003). It is a vital standard for ensuring sustainable practices. The implementation of this basic environmental standard facilitates the deployment of further standards for life cycle assessment and so on.

Six steps in compliance with the ISO 14001 standard are (Bansal & Hunter, 2003)

1. Devise an environmental action plan.
2. Recognize the organization's tasks that interface with the environment.
3. Recognize legislative/regulatory prerequisites.
4. Recognize the organization's priorities and set goals for minimizing its environmental impacts.
5. Alter the organization's structure to comply with those goals, such as responsibility allocation, training, communication and documentation.
6. Verify and check the EMS.

ISO 14001 stipulations are in line with conventional management principles (Boiral & Sala, 1998; Pun & Hui, 2001).

6.1.1 Clauses of ISO 14001

The ISO 14001 standard includes EMS specifications and 17 clauses, or general stipulations, in five groups. The requirements imply general outcomes, but do not provide specific methods an organization needs to deploy.

A summary of the 17 ISO 14001 clauses (MacDonald, 2005) are shown in Table 6.1.

TABLE 6.1

Details of 17 ISO 14001 Clauses

ISO 14001 Clauses				
4.1 Environmental policy	**4.2 Planning**	**4.3 Implementation and operation**	**4.4 Checking and corrective action**	**4.5 Management review**
—	4.2.1 Environmental aspects 4.2.2 Legal and other requirements 4.2.3 Objectives and targets 4.2.4 Environmental management program(s)	4.3.1 Structure and responsibility 4.3.2 Training, awareness and competence 4.3.3 Communication 4.3.4 EMS documentation 4.3.5 Document control 4.3.6 Operational control 4.3.7 Emergency preparedness and response	4.4.1 Monitoring and measurement 4.4.2 Non-conformance and corrective and preventive action 4.4.3 Records 4.4.4 EMS audit	—

6.1.2 Advantages of Implementing ISO 14000

The complete ISO 14000 family contributes management tools for companies to govern their environmental issues and to enhance their environmental performance. In total, these tools can contribute vital tangible economic benefits and other advantages such as (Poksinska et al., 2003; Zeng et al., 2005; Christini et al., 2004; Tarí et al., 2012)

- Minimal resource utilization
- Minimal energy usage
- Enhanced process effectiveness
- Enhanced customer satisfaction
- Enhanced employee morale
- Enhanced market share
- Cost reduction
- Enhanced productivity
- Enhanced on-time delivery to customers
- Enhanced environmental focus
- Improved profitability
- Improved corporate image
- Improved relationships with suppliers

6.1.3 Certain ISO 14000 Series Standards for sustainability (Source: ISO 2002) (Christini et al., 2004).

14000 Guide to Environmental Management Principles, Systems, and Supporting Techniques

14040/44 Life Cycle Assessment: General Principles and Practices

6.1.4 Disadvantages of EMS

- EMS deployment requires higher human and financial resources than anticipated (Hillary, 2004).
- Problems fulfilling stakeholders' needs.
- Failure to smoothly amalgamate an EMS into a firm's quality management system (QMS).

6.2 PAS 2050 Standard

PAS 2050 is a publicly available specification for evaluating the greenhouse gas (GHG) emissions of a product over its life cycle. The evaluation approach was validated with firms across various product types, addressing several sectors.

6.2.1 Need for PAS 2050

During recent times, environmental issues impose vital stress on human development (Plassmann et al., 2010). With growing public consciousness and carbon emissions, policymakers and organizations are exploring ways to standardize measures to minimize GHG emissions (Liu et al., 2016).

Fossil fuels emit more carbon dioxide (CO_2) upon combustion and contribute more to the rising CO_2 levels in the atmosphere creating global warming (Appleby et al., 2010).

6.2.2 Phases of PAS 2050

Steps involved in the PAS 2050 standard (Appleby et al., 2010) are shown in Figure 6.1.

1. Derive a process map that captures energy flows across the product recovery approach.
2. Verify boundaries and prioritization to determine the extent of the footprint and which parts are important.
3. Data gathering about waste, energy, transport emissions and so on.
4. Execute computation.
5. Assess results and validate uncertainties.

6.2.3 Organizational Benefits

The benefits of PAS 2050 include (Specification, 2008; BSI, 2008):

- Internal evaluation of product life cycle GHG emissions
- Assessment of product configuration choices, operational and sourcing alternatives, etc. in line with their impact on product GHG emissions
- Corporate responsibility reporting enabled

6.2.4 Concept of Carbon Footprint

Carbon footprint is a term that explains the quantity of GHG emissions generated by a specific task, and a path for companies and individuals to evaluate their contribution to climate change (BSI, 2008).

Variants of a carbon footprint include personal, product and organizational. The details can be found in Gao et al. (2013).

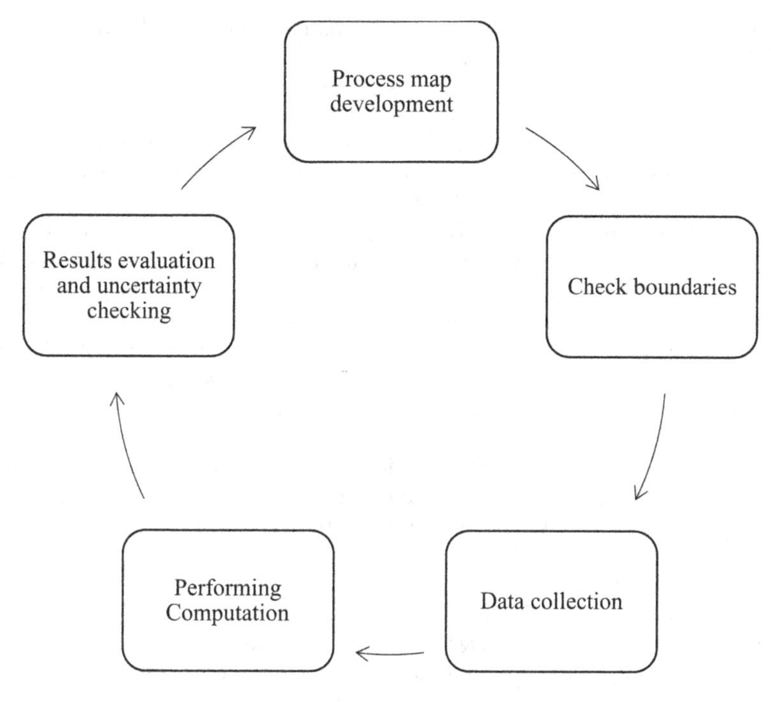

FIGURE 6.1
PAS 2050 stages.

6.3 Summary

The chapter provides readers with the details of sustainable manufacturing standards ISO 14001 EMS and PAS 2050. Scope, steps and benefits of the PAS 2050 standard are discussed. The concept of a carbon footprint and its variants are presented.

References

Appleby, M.R., Buckley, A.B., Lambert, C.G., & Rennie, A.E. (2010, January). A comparison of carbon footprint calculations for end of life product recovery methods using PAS 2050. In *ASME 2010 10th biennial conference on engineering systems design and analysis* (pp. 705–713). American Society of Mechanical Engineers Digital Collection, Istanbul, Turkey.

Bansal, P., & Hunter, T. (2003). Strategic explanations for the early adoption of ISO 14001. *Journal of Business Ethics, 46*(3), 289–299.

Boiral, O., & Sala, J.M. (1998). Environmental management: Should industry adopt ISO 14001? *Business Horizons, 41*(1), 57–64.

BSI. (2008). Guide to PAS 2050: How to assess the carbon footprint of goods and services.

Christini, G., Fetsko, M., & Hendrickson, C. (2004). Environmental management systems and ISO 14001 certification for construction firms. *Journal of Construction Engineering and Management, 130*(3), 330–336.

Gao, T., Liu, Q., & Wang, J. (2014). A comparative study of carbon footprint and assessment standards. *International Journal of Low-Carbon Technologies, 9*(3), 237–243.

Hillary, R. (2004). Environmental management systems and the smaller enterprise. *Journal of Cleaner Production, 12*(6), 561–569.

Liu, T., Wang, Q., & Su, B. (2016). A review of carbon labeling: Standards, implementation, and impact. *Renewable and Sustainable Energy Reviews, 53*, 68–79.

MacDonald, J.P. (2005). Strategic sustainable development using the ISO 14001 standard. *Journal of Cleaner Production, 13*(6), 631–643.

Plassmann, K., Norton, A., Attarzadeh, N., Jensen, M.P., Brenton, P., & Edwards-Jones, G. (2010). Methodological complexities of product carbon footprinting: A sensitivity analysis of key variables in a developing country context. *Environmental Science and Policy, 13*(5), 393–404.

Poksinska, B., Dahlgaard, J.J., & Eklund, J.A. (2003). Implementing ISO 14000 in Sweden: motives, benefits and comparisons with ISO 9000. *International Journal of Quality and Reliability Management, 20*(5), 585–606.

Pun, K.F., & Hui, I.K. (2001). An analytical hierarchy process assessment of the ISO 14001 environmental management system. *Integrated Manufacturing Systems, 12*(5), 333–345.

Specification (PAS) (2008). Specification for the assessment of the life cycle greenhouse gas emissions of goods and services. *BSI British Standards. ISBN, 978*, 580.

Tarí, J.J., Molina-Azorín, J.F., & Heras, I. (2012). Benefits of the ISO 9001 and ISO 14001 standards: A literature review. *Journal of Industrial Engineering and Management (JIEM), 5*(2), 297–322.

Zeng, S.X., Tam, C.M., Tam, V.W., & Deng, Z.M. (2005). Towards implementation of ISO 14001 environmental management systems in selected industries in China. *Journal of Cleaner Production, 13*(7), 645–656.

7

Product Sustainability and Risk–Benefit Assessment, and Corporate Social Responsibility

This chapter deals with the fundamentals of product sustainability and risk–benefit assessment with a case study. The concept of corporate social responsibility (CSR) as well its drivers are discussed.

7.1 Product Sustainability and Risk–Benefit Assessment

It is a generalized tool for calculating the Environmental Performance Index (EPI) of a product. EPI is calculated by evaluating the weighted sum of various elements of the environment that starts from material extraction, processing, design, manufacturing, transportation, use phase and finally the end-of-life (EoL) phase. The designer first defines the influencing factors in each element of the environment. The next step is to collect and record the scores of all defined factors under each element of the environment. The final step is to calculate EPI by summing the scores of all factors under each element of the environment. The equation used for calculating the EPI for any element is (Vinodh & Jayakrishna, 2013)

$$EPI_i = \sum_{j=1}^{n} \left(IF_{ij} \times L_{ij} \right) \tag{7.1}$$

where

EPI_i is the Environmental Performance Index of the ith element of the environment.

IF_{ij} is the score of the jth influencing factor under the ith element of the environment.

L_{ij} is the level of the jth influencing factor under the ith element of the environment.

The data for the score and level of each influencing factor is collected from experts. The data for the influencing factor score is on a 1 to 5 scale. Level data may be low, medium or high.

After collecting data for each factor, the factor is grouped under a risk or benefit category. For each factor, the EPI of benefits are added together and the EPI of risks are added together. Then the difference between the EPI benefits and EPI risks is the EPI of a particular element of the environment. Adding all EPIs of the element will give the product life cycle EPI (Vinodh & Jayakrishna, 2013):

$$EPI_{plc} = \left|\Sigma\left[Benefit - Risk\right]\right| \tag{7.2}$$

7.2 Risk–Benefit Assessment Case

Table 7.1 shows the product life cycle phases, criterion and corresponding risk/benefit score.

Total risk score = Summation of risk score of all phases

Total risk score = 1.5 + 2.5 + 0.5 + 1.25 + 1.5 + 3.25 + 3

Total risk score = 13.5

Total benefit score = Summation of benefit score of all phases

Total benefit score = 2.75 + 3 + 7 + 3.25 + 5 + 4 + 6

Total benefit score = 31

$$EPI_{plc}\left|\Sigma\left[Benefit - Risk\right]\right|$$

$$EPI_{plc} = 31 - 13.5 = 17.5$$

The risk and benefit scores and EPI of all product life cycle phases are depicted in Figure 7.1. The EPI score of 6.5 of the design phase is higher than other phases and hence the design phase needs major focus.

7.3 Corporate Social Responsibility (CSR): Overview

Corporate social responsibility (CSR) is the concept that firms have a requirement to integrate clusters in society other than stakeholders and beyond that stipulated by law or union contract (Jones, 1980).

TABLE 7.1

Environmental Performance Index Computation Based on Risk–Benefit Assessment

			Environmental Impact				
Product Life Cycle Phase	**Criterion**	**Score (1–5)**	**Risk**	**Benefit**	**High**	**Medium**	**Low**
Raw material extraction (RME)	Evacuation	2		1	1		
	Level of hazardous air resource	1	1			0.5	
	Toxic fumes (susceptibility)	1	1		1		
	Groundwater reduction	1		1			0.25
	Hazardous content	1		1		0.5	
TRS (RME)	1.5						
TBS (RME)	2.75						
Material processing (MP)	Quantity of waste generated	2		1	1		
	Hazard impact of the waste obtained	1	1				0.25
	Energy consumed (processing)	1		1		0.5	
	Carbon footprint (CF)	2	1		1		
	Water eutrophication (WE)	1		1		0.5	
	Air acidification (AA)	1	1				0.25
TRS (MP)	2.5						
TBS (MP)	3						
Design	Material choice	2		1	1		
	Disassembly	2		1		0.5	
	Material quantity	1		1	1		
	Cost	1	1			0.5	
	Minimal number of parts	2		1	1		
	Simple and ease of design for fabrication	2		1		0.5	

(Continued)

TABLE 7.1 (CONTINUED)

Environmental Performance Index Computation Based on Risk–Benefit Assessment

Product Life Cycle Phase	Criterion	Score (1–5)	Environmental Impact				
			Risk	Benefit	High	Medium	Low
TRS (Design)	0.5						
TBS (Design)	7						
Manufacturing	CF	2	1				0.25
	WE	1	1				0.25
	AA	1	1			0.5	
	Production methods	2		1	1		
	Energy consumption	1		1	1		
	Landfill quantity	1		1			0.25
TRS (Manufacturing)	1.25						
TBS (Manufacturing)	3.25						
Distribution and sales (Distn and Sales)	Transportation	2	1			0.5	
	Supply chain	1		1	1		
	Network complexity	1		1	1		
	Quantity handled	2		1		0.5	
	Packing method and material	2		1	1		
	Material handling	1	1			0.5	
TRS (Distn and Sales)	1.5						
TBS (Distn and Sales)	5						

(Continued)

TABLE 7.1 (CONTINUED)

Environmental Performance Index Computation Based on Risk–Benefit Assessment

			Environmental Impact				
Product Life Cycle Phase	**Criterion**	**Score (1–5)**	**Risk**	**Benefit**	**High**	**Medium**	**Low**
Product use phase	Energy consumption	3	1		1		
	Waste generated	2		1		0.5	
	CF	3		1		0.5	
	WE	1		1		0.5	
	AA	2		1		0.5	
	Noise level	1	1				0.25
TRS (Product use)	3.25						
TBS (Product use)	4						
End of Life (EoL)	Recycling characteristics	3		1	1		
	Remanufacturing characteristics	3		1	1		
	Reconditioning characteristics	2	1				0.25
	Repair alternatives	1	1				0.25
	Recovery options	1	1				0.25
	Reuse possibilities	2	1		1		
TRS (EoL)	3						
TBS (EoL)	6						

Note: TRS denotes the total risk score and TBS denotes the total benefit score.

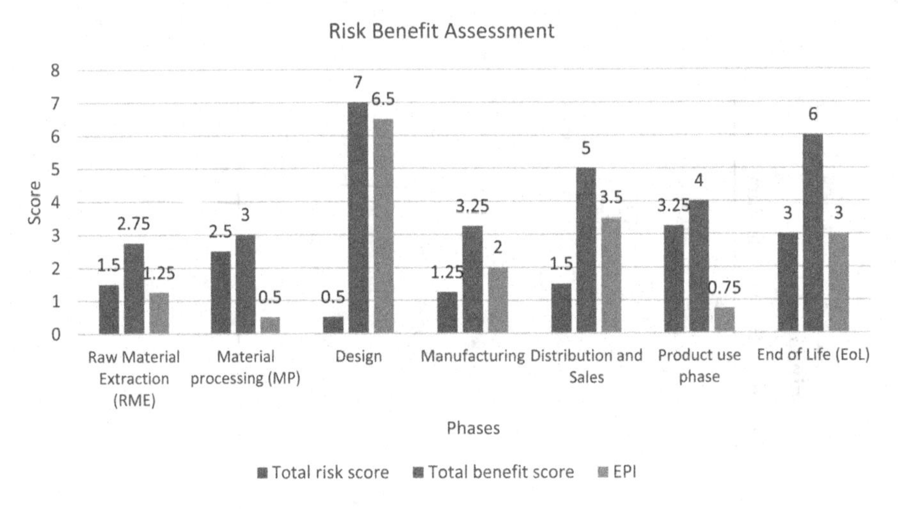

FIGURE 7.1
Risk score, benefit score and EPI for product life cycle phases.

7.3.1 Definitions

- The European Commission (2010) definition of CSR is "a concept whereby companies integrate social and environmental concerns in their business operations and in their interaction with their stakeholders on a voluntary basis".
- Definition of CSR by the World Business Council for Sustainable Development (WBCSD): "The continuing commitment by business to behave ethically and contribute to sustainable economic development while improving the quality of life of the workforce and their families as well as of the local community and society at large" (Khan et al., 2012).

7.3.2 Scope

- CSR pertains to the tasks of businesses, by means of their role to attaining triple bottom line (TBL) sustainability dimensions (Jenkins, 2009).
- CSR motivates a firm to contribute to society as it facilitates the organization to deploy CSR initiatives in line with the organization's vision and mission (Othman & Abdellatif, 2011).
- CSR promotes green production practices, namely energy consumption, emission reduction, usage of recycled materials and minimizing packing materials (Sprinkle & Maines, 2010).
- Vendors and customers are involved with their firms in environmental initiatives (Sprinkle & Maines, 2010).

7.3.3 Need for CSR are (Sprinkle & Maines, 2010)

- Firms agree that CSR enables employee recruitment and retaining.
- CSR may attract consumers to purchase a firm's goods.
- Organizations concentrate on environmental issues that lead to a reduction in manufacturing costs (Sprinkle & Maines, 2010).

7.3.4 Benefits of CSR are (Sprinkle & Maines, 2010)

- Enhanced cash inflow to the company.
- Positive reputation on social media for the organization.
- Means for attraction, motivation and retaining of skills.
- Efficiency improvement and cost benefits in the value chain.
- Firms might experience a relationship between their CSR initiatives and consumers' buying patterns (Sprinkle & Maines, 2010).

7.3.4.1 Business Benefits of CSR are (Drews, 2010)

- Increase in revenue.
- Cost reduction.
- Risk minimization.
- Enhanced brand value.
- Enhanced access to capital.
- Enhanced customer attraction and retention.
- Enhanced image (Drews, 2010).

7.4 Drivers of CSR

The drivers of CSR are presented in four categories – market, societal, government and globalization – in Table 7.2 (Tuzzolino & Armandi, 1981; Govindan et al., 2014).

7.5 Summary

This chapter details the procedure of product sustainability and risk–benefit assessment with computation details. Details of various criteria, risk score, benefit score and Environmental Performance Index

TABLE 7.2

Drivers of CSR

Driver Category	Drivers
Market	• Goal alignment • Reputation • Company's image • Investors
Social	• Job enhancement • Business ethics • Regulations for workers • Business associations for CSR • Employee relations
Government	• External (community and government relations) • Government regulations • Emissions reduction • Subsidy
Globalization	• Financial ratios • Strategic advantage • Competitive position • Business authenticity

calculation for different product life cycle phases are discussed. Scope, need, benefits and drivers of CSR are also covered.

References

Drews, M. (2010). Measuring the business and societal benefits of corporate responsibility. *Corporate Governance, 10*(4), 421–431.

European Commission. (2010). Corporate social responsibility (CSR). Retrieved from http://ec.europa.eu/enterprise/policies/sustainable-business/corporate-social-responsibility/index_en.htm.

Govindan, K., Kannan, D., & Shankar, K.M. (2014). Evaluating the drivers of corporate social responsibility in the mining industry with multi-criteria approach: A multi-stakeholder perspective. *Journal of Cleaner Production, 84*, 214–232.

Jenkins, H. (2009). A 'business opportunity' model of corporate social responsibility for small- and medium-sized enterprises. *Business Ethics: A European Review, 18*(1), 21–36.

Jones, T.M. (1980). Corporate social responsibility revisited, redefined. *California Management Review, 22*(3), 59–67.

Khan, M.T., Khan, N.A., Ahmed, S., & Ali, M. (2012). Corporate social responsibility (CSR)–definition, concepts and scope. *Universal Journal of Management and Social Sciences, 2*(7), 41–52.

Othman, A., & Abdellatif, M. (2011). Partnership for integrating the corporate social responsibility of project stakeholders towards affordable housing development: A South African perspective. *Journal of Engineering, Design and Technology, 9*(3), 273–295.
Sprinkle, G.B., & Maines, L.A. (2010). The benefits and costs of corporate social responsibility. *Business Horizons, 53*(5), 445–453.
Tuzzolino, F., & Armandi, B.R. (1981). A need-hierarchy framework for assessing corporate social responsibility. *Academy of Management Review, 6*(1), 21–28.
Vinodh, S., & Jayakrishna, K. (2013). Assessment of product sustainability and the associated risk/benefits for an automotive organisation. *The International Journal of Advanced Manufacturing Technology, 66*(5–8), 733–740.

8

Sustainability Assessment

This chapter presents sustainability indicators based on a triple bottom line (TBL) perspective and details sustainability assessment models reported in the literature. Multi-grade fuzzy and fuzzy logic assessment methods are detailed for sustainability assessment.

8.1 Sustainability Indicators (Environment, Economy and Society Based)

Tables 8.1, 8.2 and 8.3 present sustainability indicators (environment, economy and society). The indicators are presented with their corresponding literature.

8.2 Sustainability Assessment Models

The details of sustainability assessment models are presented in this section. Seven models are derived from the literature and their details are discussed in this section.

8.2.1 A Generic Framework for Sustainability Assessment of Manufacturing Processes

A framework was contributed by Saad et al. (2019) for assessing sustainability of manufacturing processes. The objective was to develop a new and detailed architecture for sustainability assessment of manufacturing processes in line with TBL sustainability dimensions. The architecture integrates objective and subjective weighting methods to minimize uncertainty with subjective weighting. Also, interaction among various indicators were captured using multi-criteria decision making (MCDM) methods.

TABLE 8.1

Sustainability Indicators (Environment Based)

Environment	Company's reputation	Junior et al. (2018)
	Compliance with the environmental legislation	Junior et al. (2018)
	Resource consumption	Saad et al. (2019), Sabaghi et al. (2016), Vinodh (2011), Singh et al. (2014)
	Mineral and energy resources	Vinodh (2011)
	Air pollution	Sabaghi et al. (2016)
	Soil pollution	Sabaghi et al. (2016)
	Material utilization	Shuaib et al. (2014)
	Regulations and certification	Shuaib et al. (2014)
	Energy from renewable sources	Shuaib et al. (2014)
	Energy from non-renewable sources	Shuaib et al. (2014)
	Energy efficiency	Shuaib et al. (2014)
	Water use	Shuaib et al. (2014)
	Solid waste	Shuaib et al. (2014)
	Liquid waste	Shuaib et al. (2014)
	Product EoL	Shuaib et al. (2014)
	Reused material ratio	Singh et al. (2014)
	Recyclable material ratio	Singh et al. (2014)
	Non-renewable material ratio	Singh et al. (2014)
	Hazardous material ratio	Singh et al. (2014)
	Waste material ratio	Singh et al. (2014)
	Renewable energy ratio	Singh et al. (2014)
	Waste water ratio	Singh et al. (2014)

TABLE 8.2

Sustainability Indicators (Economy Based)

Economy	Productivity	Junior et al. (2018)
	Profitability	Junior et al. (2018), Saad et al. (2019)
	Salary and benefits	Junior et al. (2018)
	Turnover	Junior et al. (2018)
	Market share	Junior et al. (2018)
	Investment	Saad et al. (2019), Sabaghi et al. (2016)
	Potential financial benefits	Vinodh (2011), Sabaghi et al. (2016)
	Trading opportunities	Vinodh (2011)

8.2.2 Integrated Sustainability Assessment Framework

Three different frameworks for social, economic and environment criteria were developed by Bhanot et al. (2016). The economic framework includes 18 economic indicators for different aspects of the economy and are grouped in four categories: manufacturing cost, machining performance,

TABLE 8.3

Sustainability Indicators (Society Based)

Society	Job opportunities	Vinodh (2011)
	Employment compensation	Vinodh (2011)
	Research and development	Vinodh (2011)
	Security	Vinodh (2011)
	Social cohesion	Vinodh (2011), Junior et al. (2018)
	Regulatory and public services	Vinodh (2011)
	Stakeholder participation	Vinodh (2011)
	Macro social performance	Vinodh (2011)
	Ethics and transparency	Junior et al. (2018)
	Compliance with social legislation	Junior et al. (2018)
	Skill level of worker	Bhanot et al. (2016)
	Punctuality to work	Bhanot et al. (2016)
	Job satisfaction level	Bhanot et al. (2016)
	Conducive working environment	Bhanot et al. (2016)
	Extent of government support	Bhanot et al. (2016)
	Occupational health	Sabaghi et al. (2016), Vinodh (2011),
	Workplace environment	Sabaghi et al. (2016)
	Safety risk	Sabaghi et al. (2016), Vinodh (2011)
	Product safety and health impact	Shuaib et al. (2014)
	Employee turnover ratio	Singh et al. (2014)
	Training hours/employee	Singh et al. (2014)

production efficiency and process improvement. The social framework consists of 24 indicators focused on workforce training, health and safety. Twenty-four social indicators are grouped into four categories: health issues, safety issues, labour issues and workforce training. The environmental framework consists of 28 indicators pertaining to various aspects of the environment such as waste management and resource consumption. The defined 28 indicators for environment are grouped in five categories, namely water consumption, energy consumption, material type, pollution and waste, and government rules and regulations. The developed framework was comprehensive and includes 70 indicators from all three perspectives of the TBL approach. All 70 indicators were related to the manufacturing sector.

8.2.3 Three-Level Conceptual Model for Sustainability Assessment

A three-layer sustainability assessment framework was developed by Vinodh (2011). The first layer consists of sustainability enablers, namely environment, economy and society. These three sustainability enablers are further divided into sustainability criteria which form the second layer of the framework. The sustainability criteria in the second level are then further divided into sustainability attributes which form the third layer. This framework includes 3 sustainability enablers, followed by 12 sustainability criteria and 37 sustainability attributes in the first, second and third levels, respectively. The developed framework was comprehensive and based on the 3Ps (profitability, people and planet).

8.2.4 Sustainability Assessment Model for SMEs

A two-level framework was developed by Singh et al. (2014) consisting of three categories and 21 indicators for small and medium-sized enterprises (SMEs). The three categories are the economic, environmental and social aspects of sustainability which include four, twelve and five indicators, respectively. With reference to the case reported in the study, the following indicators were found relevant. The economic category includes profit and quality aspects. The environment category includes material intensity, pollutants and waste, and the social category includes employee satisfaction and community projects. The framework was developed with the focus on manufacturing SMEs.

8.2.5 A Metrics-Based Framework to Assess Total Life Cycle Sustainability of Manufactured Products

A four-layer framework for evaluating sustainability throughout a product's lifetime was developed by Shuaib et al. (2014). The four layers are, in order, subindex, category, subcategory and individual metrics. The first layer, subindex, consists of three sustainability dimensions namely, economy, environment and society. Category, at the second layer, includes 14 major factors of sustainability. Subcluster, in the third layer, includes 45 more specific clusters in a detailed manner. At the fourth level, individual metrics include more than 80 evaluation parameters for assessing the sustainability index of the product.

8.2.6 Sustainability Assessment Using Fuzzy Inference Technique

A sustainability assessment framework in the form of hierarchy levels was proposed by Sabaghi et al. (2016). The developed framework consists of 55 influencing factors, 9 primary criteria at level I and 12 secondary criteria at level II, 3 TBL criteria at level III, and finally an overall sustainability index at level IV. The developed framework evaluates the overall sustainability index by considering all three aspects of sustainability.

8.2.7 Sustainability Evaluation Model for Manufacturing Systems

A sustainability evaluation model was developed in integration with the balanced scorecard (BSC) perspective by Junior et al. (2018). The developed model was in the form of a matrix, whereby rows contained the three dimensions of TBL (environment, economy and society) and columns contain the four perspectives of the BSC (learning and growth, process, market, and financial). The cells of the matrix consist of 12 sustainability indicators. The unique aspect of the developed model is that it was based on correlation between the TBL dimensions and 4 perspectives of BSC, which resulted in 12 correlations, also termed 12 indicators.

8.3 Multi-Grade Fuzzy Assessment Approach

The multi-grade fuzzy (MGF) method is applied to evaluate the sustainability performance index (Vinodh et al., 2010). In MGF methodology, the sustainability performance index is computed using the following equation:

$$I = W * R \tag{8.1}$$

where

I is the sustainability performance index.
W is the overall weight of sustainability indicators.
R is the overall rating of sustainability indicators.

In this method, the sustainability performance index is divided in five grades:

$$I = \{x_1, x_2, x_3, x_4, x_5\} \; or \; I = \{10, 8, 6, 4, 2\}$$

The range 10–8 shows 'very good sustainable performance', 8–6 represents 'good sustainable performance', 6–4 indicates 'moderate sustainable performance', 4–2 represents 'low sustainable performance', and less than 2 represents 'very low sustainable performance' (Yang & Li, 2002).

To implement the MGF method, a framework needs to be developed for assessment of the sustainable performance level (Vinodh & Chintha, 2011). The framework can be in the form of two or three layers. A three-layer framework format is shown in Table 8.4, where weights are group weights and ratings are individual ratings.

Primary calculation:

$$I_{11} = R_{11} * W_{11} \tag{8.2}$$

TABLE 8.4
MGF Approach (Three-Layer Framework)

Enabler	W_1	Criteria	W_{11}	Attribute	R_1	R_2	R_3	W_{111}
I_1		I_{11}		I_{111}				

Secondary calculation:

$$I_1 = R_1 * W_1 \tag{8.3}$$

Tertiary calculation:

$$I = R * W \tag{8.4}$$

The overall sustainability performance is calculated by solving the developed framework using the preceding computation.

8.4 Fuzzy Logic Assessment Approach

Fuzzy logic (FL) is a methodology aimed at analyzing the sustainable performance of firms. The steps involved in fuzzy logic are as follows.

Step 1: Identify the criteria and attributes based on which assessment needs to be done.

Step 2: Develop a framework for sustainability assessment, consisting of three layers. The format of a three-layer framework used in the FL approach is shown in Table 8.5.

TABLE 8.5
FL Approach (Three-Layer Framework)

Enabler	W_1	Criteria	W_{11}	Attribute	R_1	R_2	R_3	W_{111}
I_1		I_{11}		I_{111}				

Step 3: Select and define the appropriate linguistic scale considered for gathering data for performance ratings and group weights of sustainability indicators. The standard linguistic scale and corresponding fuzzy numbers for weights are given next (Lin et al., 2006):

Very low (VL), (0, 0.05, 0.1); low (L), (0.1, 0.2, 0.3); fairly low (FL), (0.2, 0.35, 0.5); medium (M), (0.3, 0.5, 0.7); fairly high (FH), (0.5, 0.65, 0.8); high (H), (0.7, 0.8, 0.9); very high (VH), (0.8, 0.95, 1).

The standard linguistic scale and associated fuzzy numbers for performance ratings of sustainability indicators are given in the following:

Worst (W), (0, 0.5, 1.5); very poor (VP), (1, 2, 3); poor (P), (2, 3.5, 5); fair (F), (3, 5, 7); good (G), (5, 6.5, 8); very good (VG), (7, 8, 9); excellent (E), (8, 9.5, 10).

Step 4: Collect and measure data pertaining to performance ratings and weights using the linguistic scale.

Step 5: The collected data are then approximated using fuzzy numbers.

Step 6: Based on the equations, the primary, secondary and tertiary calculations need to be done.

Primary calculation:

The sustainability index of the *j*th criteria of the *i*th enabler is computed using Equation 8.5:

$$SI_{ij} = \frac{\sum_{i=1}^{n} R_{ijk} * W_{ijk}}{\sum_{i=1}^{n} W_{ijk}} \tag{8.5}$$

where

SI_{ij} is the sustainability index of the *j*th criteria pertaining to the *i*th enabler (Lin et al., 2006).

R_{ijk} is the rating for each sustainability indicator.

W_{ijk} is the weight for each sustainability indicator.

n is the number of sustainability indicators.

Secondary calculation:

The sustainability index of the *i*th enabler is computed using the following equation:

$$SI_{i} = \frac{\sum_{i=1}^{m} R_{ij} * W_{ij}}{\sum_{i=1}^{m} W_{ij}} \tag{8.6}$$

where

SI_i is the sustainability index of the *i*th enabler (Lin et al., 2006).
R_{ij} is the rating for each sustainability criteria.
W_{ij} is the weight for each sustainability criteria.
m is the number of sustainability criteria.

Tertiary calculation:

The fuzzy logic sustainability index is calculated using the following equation:

$$FLSI = \frac{\sum_{i=1}^{n} R_i * W_i}{\sum_{i=1}^{n} W_i} \tag{8.7}$$

where

FLSI is the Fuzzy Logic Sustainability Index (Lin et al., 2006).
R_i is the rating for sustainability enablers.
W_i is the weight for sustainability enablers.
n is the number of sustainability enablers.

Step 7: Based on step 6, the overall fuzzy logic index is calculated, which shows the sustainability level of the case firm.

Step 8: The Euclidean distance approach is used to benchmark the calculated sustainability index with the standard scale to find the implementation level.

In the Euclidean distance approach, the sustainability level of an organization can be evaluated by identifying the distance between the computed sustainability index and standard sustainability levels. The standard sustainability levels (Lin et al., 2006) are slightly sustainable (SS), fairly sustainable (FS), sustainable (S), very sustainable (VS) and extremely sustainable (ES).

The equation for calculating Euclidean distance is

$$D(FLSI, SI_i) = \left\{ \sum \left(f_{FLI}(x) - f_{SIi}(x) \right)^2 \right\}^{1/2} \tag{8.8}$$

where

$f_{FLI}(x)$ is the fuzzy number for the FLSI.
$f_{SIi}(x)$ is the fuzzy number for standard sustainability levels.

Step 9: The Fuzzy Performance Importance Index (FPII) needs to be computed to recognize the weaker areas of sustainability implementation so that focus can be given to identified weaker areas.

The equation for calculating the FPII is

$$FPII = W'_{ijk} * R_{ijk} \tag{8.9}$$

where

W'_{ijk} is the complementary weights of sustainability indicators.
R_{ijk} is the performance ratings of sustainability indicators.

In comparison to the MGF approach, the fuzzy logic approach could better deal with uncertainty. Also identification and analysis of weaker areas can be done more effectively.

8.5 Summary

This chapter deals with sustainability indicators (environment, economy and society based) identified in the literature. Seven sustainability assessment models were reviewed in terms of focus and indicators were discussed. The multi-grade fuzzy and fuzzy logic assessment approaches are discussed from the viewpoint of sustainability evaluation. The computational steps of sustainability index calculation are presented.

References

Bhanot, N., Rao, P.V., & Deshmukh, S.G. (2016). An integrated sustainability assessment framework: A case of turning process. *Clean Technologies and Environmental Policy, 18*(5), 1475–1513.

Junior, A.N., de Oliveira, M.C., & Helleno, A.L. (2018). Sustainability evaluation model for manufacturing systems based on the correlation between triple bottom line dimensions and balanced scorecard perspectives. *Journal of Cleaner Production, 190*, 84–93.

Lin, C.T., Chiu, H., & Tseng, Y.H. (2006). Agility evaluation using fuzzy logic. *International Journal of Production Economics, 101*(2), 353–368.

Saad, M.H., Nazzal, M.A., & Darras, B.M. (2019). A general framework for sustainability assessment of manufacturing processes. *Ecological Indicators, 97*, 211–224.

Sabaghi, M., Mascle, C., Baptiste, P., & Rostamzadeh, R. (2016). Sustainability assessment using fuzzy-inference technique (SAFT): A methodology toward green products. *Expert Systems with Applications, 56*, 69–79.

Shuaib, M., Seevers, D., Zhang, X., Badurdeen, F., Rouch, K.E., & Jawahir, I.S. (2014). Product sustainability index (ProdSI): A metrics-based framework to evaluate the total life cycle sustainability of manufactured products. *Journal of Industrial Ecology, 18*(4), 491–507.

Singh, S., Olugu, E.U., & Fallahpour, A. (2014). Fuzzy-based sustainable manufacturing assessment model for SMEs. *Clean Technologies and Environmental Policy, 16*(5), 847–860.

Vinodh, S. (2011). Assessment of sustainability using multi-grade fuzzy approach. *Clean Technologies and Environmental Policy, 13*(3), 509–515.

Vinodh, S., & Chintha, S.K. (2011). Leanness assessment using multi-grade fuzzy approach. *International Journal of Production Research, 49*(2), 431–445.

Vinodh, S., Devadasan, S.R., Vasudeva Reddy, B., & Ravichand, K. (2010). Agility index measurement using multi-grade fuzzy approach integrated in a 20 criteria agile model. *International Journal of Production Research, 48*(23), 7159–7176.

Yang, S.L., & Li, T.F. (2002). Agility evaluation of mass customization product manufacturing. *Journal of Materials Processing Technology, 129*(1–3), 640–644.

9

Software Modules for Life Cycle Assessment (LCA) and Sustainable Manufacturing

9.1 Product-Based Sustainability Analysis Module

9.1.1 Sustainability Xpress: SolidWorks Sustainability Analysis Module

Sustainability Xpress is a SolidWorks add-in that enables users to generate more sustainable designs with reference to material type, manufacturing process, material use and environmental impacts (Dassault Systèmes, 2010). The data used is in line with PE International's database and its GaBi software (Systèmes, 2010).

9.1.2 Need

A life cycle assessment (LCA) method that integrates computer-aided design (CAD) is a vital tool to benchmark various design solutions and to substantiate real environmental advantage. The module is incorporated with simplified LCA that enables design researchers to predict sustainability in initial design phases (Morbidoni et al., 2011).

9.1.3 Scope

The module is a product-based LCA tool that assesses product designs from an environmental perspective to facilitate benchmarking of various design solutions connected with material and conversion process selection (Morbidoni et al., 2011).

Inputs to this module include a CAD model of the part, material, manufacturing process, manufacturing and user location. The four impacts are carbon footprint, eutrophication, acidification and energy. Design engineers can control these values by choosing various transportation modes, manufacturer location, manufacturing method(s) and materials, which facilitate them to develop eco-friendly products. This tool provides a way to assess a component's environmental impact through four benchmark parameters (Rodriguez & Christensen, 2014).

The module provides designers access to environmental databases so as to select material and manufacturing processes in such a way to make relevant decisions with regard to environmental impacts and developing sustainable products (Paudel & Fraser, 2013).

9.1.4 Environmental Impact

The four environmental indicators generated by the module are (Morbidoni et al., 2011)

- Carbon footprint or global warming potential (CO_2)
- Total energy consumed (MJ)
- Air acidification (SO_2)
- Water eutrophication (PO_4)

Details of the environmental impact parameters can be found in Morbidoni et al. (2011) and Dassault Systèmes (2010).

9.1.5 Steps in Sustainability Xpress

Sustainability Xpress includes four menus (Dassault Systèmes, 2010): Material, Manufacturing, Transportation and Use, and Environmental Impact. The Material menu gathers data pertaining to the part solid model. Using the material selection menu, the appropriate material is set. In the Manufacturing menu, the appropriate process is set. The Transportation and Use menu is set by selecting the desired region on the maps depicted for both parts. The Environmental Impact menu is below the Manufacturing, and Transportation and Use sections.

The input parameters for environmental analysis of the design are (Paudel & Fraser, 2013)

1. Material selection
2. Manufacturing process selection
3. Manufacturing location selection
4. Use location selection

The database includes various engineering materials enabling the designer to recognize low-impact materials to contribute to a sustainable design. A list of manufacturing processes are available for selection. The transportation cost of the products must be included and the module enables picking the manufacturing and user locations (Paudel & Fraser, 2013).

The details regarding options and menus pertaining to materials, manufacturing (process and use menus), and report generation can be found in the literature (Dassault Systèmes, 2010; Paudel & Fraser, 2013).

9.1.6 Sustainability Report

The impact of each environmental indicator is displayed in pie and bar charts. The module allows setting a baseline and performing a benchmark study. Appropriate colours are used to depict different states.

9.1.6.1 Analysis Results of a Case Study Product

The component considered in the study was a critical automotive component. The design is checked for alternate materials and manufacturing processes, and the main goal was weight reduction and minimization of environmental impacts during the product life cycle. Presently, the automotive part is made up of 1023 Steel Sheet (SS) using the turning process.

Figure 9.1 shows the environmental impact generated by the automotive part.

As an alternate material, ABS plastic with an injection moulding process was selected. As an alternative process, machining was chosen with the 1023 Steel Sheet (SS) material. In the case of ABS, steel replacement had a negative

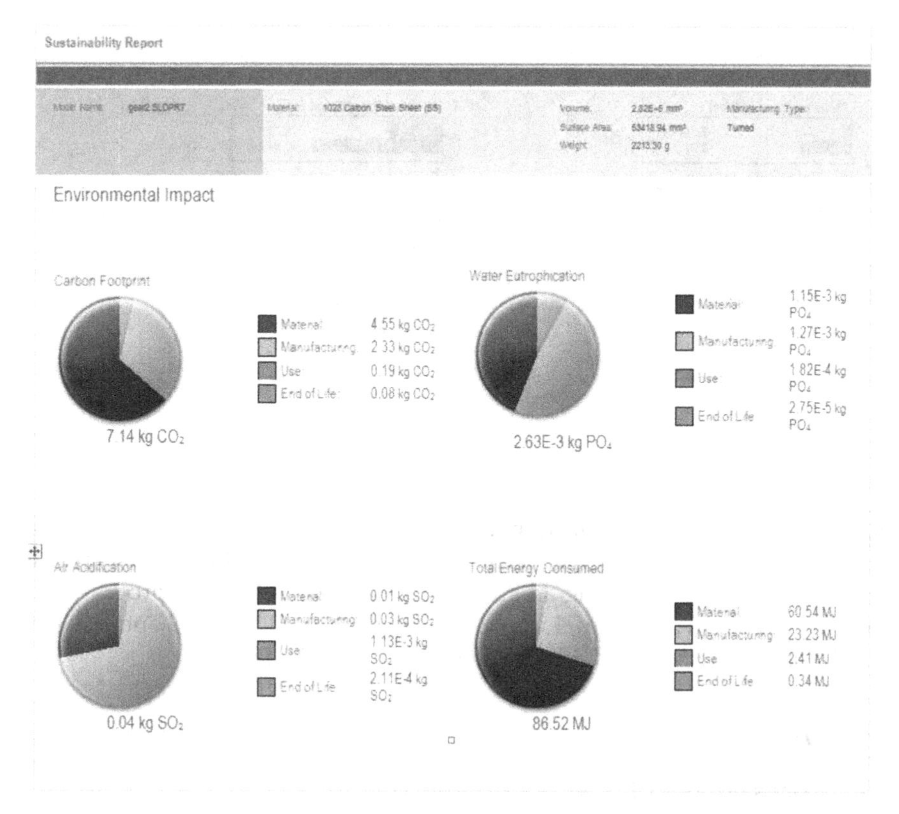

FIGURE 9.1
Sustainability assessment of a part using SolidWorks Sustainability Xpress.

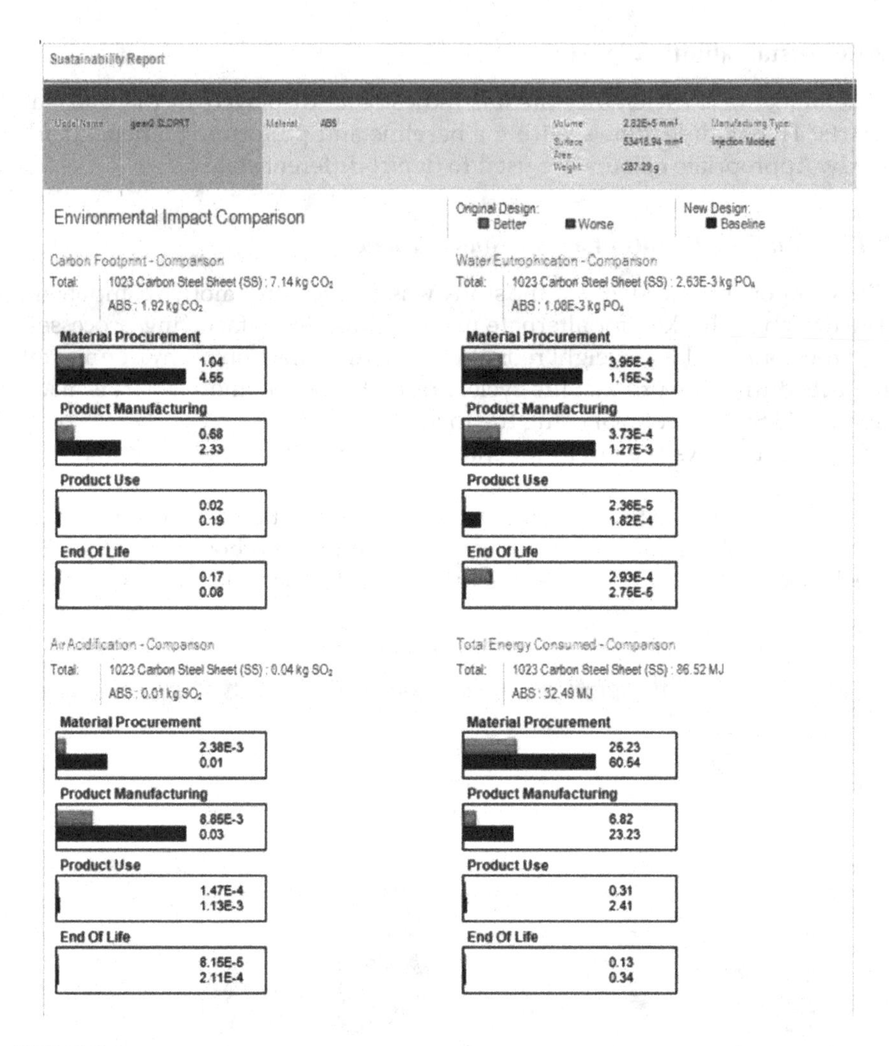

FIGURE 9.2
Sustainability report comparison by changing material.

effect on the water eutrophication level with extrusion as the manufacturing process. ABS plastics are reusable and the eutrophication effect can be drastically minimized by utilizing an appropriate effluent treatment process. The comparison of environmental impacts is depicted in Figures 9.2 and 9.3. The weight of the part was reduced by changing the material.

9.2 Process-Based LCA–GaBi Module

GaBi is a product sustainability module which helps LCA practitioners to analyze sustainable performance during the design stage itself. GaBi is a

FIGURE 9.3
Sustainability report comparison by changing manufacturing process.

developed standard module for performing LCA in any area (Gruber et al., 2016). It helps practitioners to assess the vital environmental impact of a product in its production, transportation, use and disposal phases. GaBi also facilitates carbon footprint computation, LCA, life cycle costing and benchmarking studies.

9.2.1 Features of GaBi are (Spatari et al., 2001; PE-International, 2012)

- GaBi presents a visual map showing all the process and associated flows along with system boundaries.
- User interface (ergonomically designed).

- Life cycle costing calculation (Spatari et al., 2001; PE International, 2012).

The database available in the GaBi module is biggest and complete database of wider materials (Martínez-Rocamora et al., 2016).

9.2.2 Modelling and Analysis Using GaBi

This section depicts the step-by-step procedure to perform LCA using GaBi.

9.2.2.1 Goal and Scope Definition

In this phase, the goal of the study is defined along with its boundary conditions. The goal refers to the intended purpose of doing LCA and the reasons for the study. The scope includes functions of product systems, functional unit, product system, system boundaries and so on (PE International, 2012).

9.2.2.2 Life Cycle Inventory (LCI)

The life cycle inventory (LCI) phase includes gathering and quantifying all the input and output data for a product in its entire life cycle. GaBi supports the LCI phase by providing features such as a transparent database for performing LCA and providing access to fundamental LCI data (PE International, 2012).

GaBi module has following entities for doing LCA. The entities are referred from GaBi manual, PE International, (2012).

- *Flows:* In GaBi, flow represents the material, energy or any resource flow between the processes. Flows are utilized by processes (input and output) and links between processes. The GaBi database has a pre-defined set of flows grouped by type.
- *Processes:* In GaBi, in line with flows, processes are hierarchically categorized. A set of processes exist in the database for modelling with upstream data.
- *Plans:* Plans enable assemblage of processes involved in a product system. Plan implies a process map which presents all stages and substages involved in a product system. GaBi provides nesting of plans to represent a complex system.
- *Balances:* The balance function in GaBi helps in comparing all the inputs of a system with their outputs. The balance feature of GaBi includes LCI results. The balance feature in GaBi helps in inventory analysis, impact assessment and interpretation of results.

9.2.2.3 ***Impact Assessment***

The goal of balancing is the evaluation of significant environmental impacts with classification and characterization steps.

9.2.2.4 ***Interpretation***

This step enables decision-making by analyzing results of the life cycle balance and life cycle impact assessment.

9.2.3 Procedural Steps in GaBi

The complete details of executing a project in GaBi can be found in the GaBi manual (PE International, 2012).

9.2.4 Case Study

- *Goal definition:* To evaluate the environmental impact generated due to production of aluminium part.
- *Inventory analysis:* It begins with the formulation of a flow diagram, which illustrates all related materials and energy flows. The reference flow for the configured functional unit is quantified pertaining to the flows. Initially, the flow diagrams for aluminium part production were developed.
- *Flow diagram of aluminium part production:* Extraction of raw material was done with the help of coal heat. Processing of the aluminium part was done through machining. Useful life is also included in the flow diagram along with its end-of-life scenario. Landfill is considered as the disposal option.
- *Life cycle impact assessment:* It helps to calculate the scores of impact groups formulated in the characterization model. It includes four steps:
    1. *Collection of impact groups.* In line with inventory analysis, the probable impact groups are identified using an input–output–impact (I-O-I) diagram. It enables one to map the probable impact categories for the gathered inventory.
    2. *Selection of characterization model.* The characterization model is recognized in line with the selected impact groups. The Eco-indicator 99 model is utilized in the present study.
    3. *Classification.* The contribution of inventory data towards impact categories is analyzed using the Eco-indicator 99 characterization model.

4. *Interpretation.* Figure 9.4 shows the model configured in the GaBi 5.0 LCA module for aluminium part production. In line with the developed model, sample impact scores were calculated, as depicted in Figure 9.5. The impact parameters considered were global warming potential (GWP), ozone depletion potential (ODP), human toxicity (HT), particulate matter, ionizing radiation, photochemical ozone creation potential (POCP), acidification potential (AP), eutrophication potential (EP) and ecotoxicity.

9.3 Process-Based LCA–SimaPro Module

SimaPro was developed by PRé Sustainability to efficiently utilize LCA expertise, to facilitate solid decision-making, transform products' life cycles for better and enhance a firm's reputation (PRé Consultants, 2008).

The module facilitates examination and monitoring of sustainability of products and services. It facilitates modelling and examination of complex life cycles in a meticulous way to analyze environmental impacts across life cycle phases and location of hotspots (Starostka-Patyk, 2015).

The module is used for several applications, such as sustainability reporting, carbon and water footprint assessment, product design, generation of environmental product declarations, and finding key performance indicators.

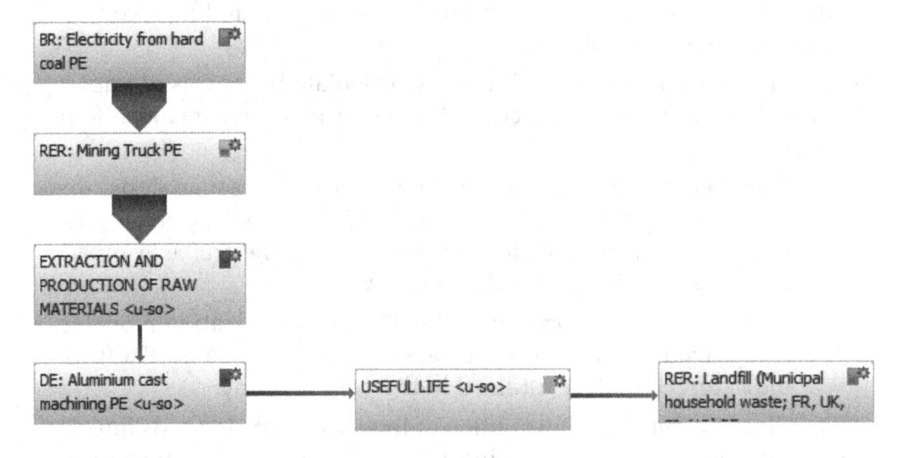

FIGURE 9.4
Plan for LCA of aluminium part production (cradle-to-grave analysis).

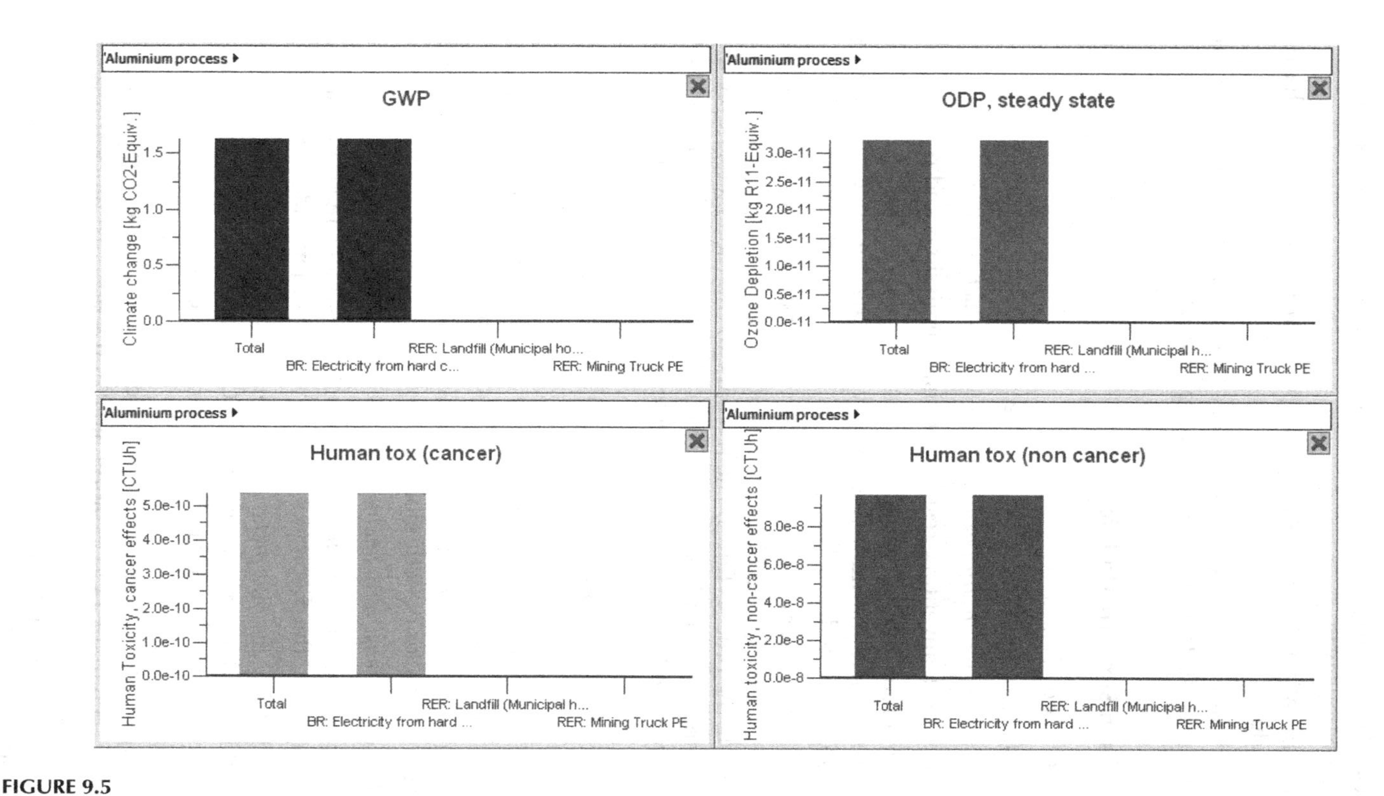

FIGURE 9.5
Results obtained after balancing for aluminium part production (GWP, ODP, HT parameters).

9.3.1 Features of SimaPro Module are (Goedkoop et al., 2016)

- Ease of modelling and analysis of complex life cycles in an effective and obvious way
- Quantification of environmental impact during all life cycle phases of products
- Identification of bottlenecks in every link of the supply chain, from raw materials to disposal (Goedkoop et al., 2016)

9.3.2 Steps in SimaPro Module

Step 1: Check goal and scope.

Step 2: Check the processes in the database.

Step 3: Examine the environmental profile of a product.

1. Life cycle inventory result with list of emissions and resources.
2. Different impact assessment steps: characterization, damage assessment, normalization and weighting are done and results are generated.

Step 4: Create a process network. The Network tab displays the network of all processes, wherein each box represents a process and arrows indicate the process flows. The red bars (or thermometers) imply the environmental load pertaining to each process, where distinction between vital and less vital processes (i.e. identify hotspots) can be shown.

Step 5: Analyze a full life cycle. Next, the full life cycle is analyzed. Two product life cycles can be set.

Step 6: Compare the production stages of two products. In this step, two products are benchmarked in terms of their production stages.

Step 7: Compare life cycles. Benchmarking of the environmental impacts of the life cycles of both models is done.

Step 8: Execute sensitivity analysis on alternative assumptions.

Step 9: Inspect or select a method. Impact assessment methods can be selected with the display of related normalization and weighting.

Step 10: Validate the interpretation section. This segment is utilized as a checklist and a framework for the LCA report. The most vital interpretation issues based on ISO 14043 can be fed here.

SimaPro organizes data entered into 'projects'. Every project has four different parts as seen in the LCA explorer: goal and scope, inventory, impact, and interpretation.

In the module, wizards can be run that include guidance for less experienced users.

Various segments in the lower segment include emissions to air, water, soil, waste flows, radiation and noise.

The details pertaining to various steps can be found in Goedkoop et al. (2016).

9.3.3 Case Study

This case presents LCA to assess environmental impacts associated with the welding process. The LCA module SimaPro based on the Eco-indicator 99 (I) approach is used. The product under consideration is an industrial component welded to a pipe. The welding process used is submerged arc welding.

Stage 1: Goal and scope definition. The goal of the study is to perform a detailed LCA for a welding process. The necessary inputs needed for performing the analysis is collected from the shop floor. The scope of LCA is limited to the welding process. The network diagram for the welding process is shown in Figure 9.6.

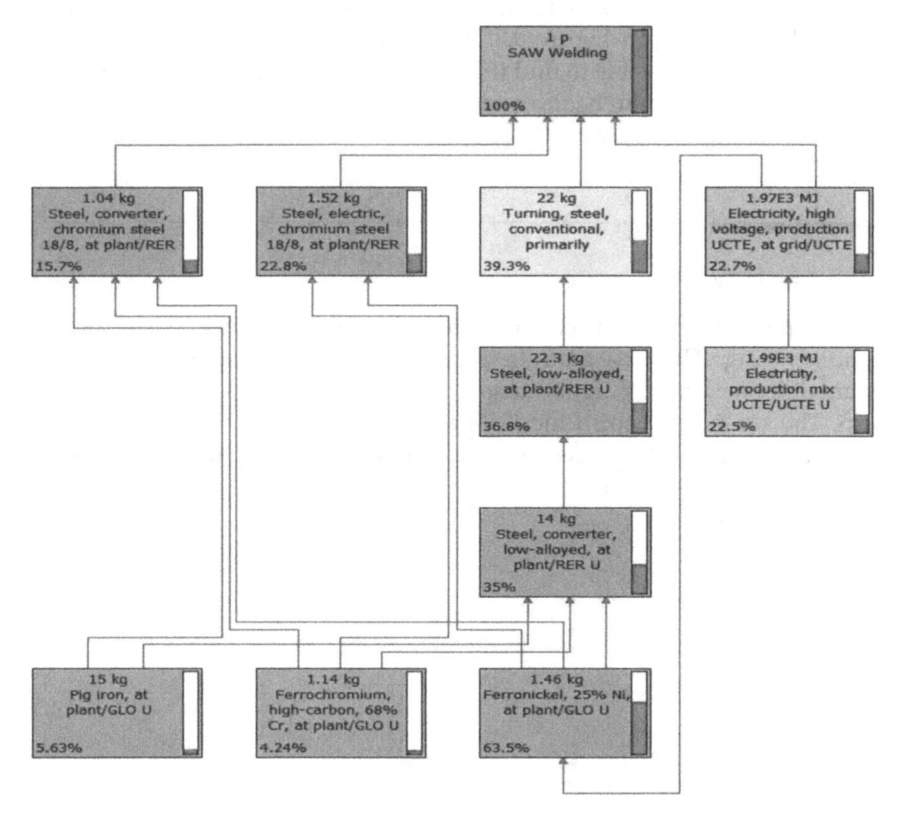

FIGURE 9.6
Network diagram of welding process in LCA module.

Stage 2: Inventory analysis stage. The needed inventory details required for performing the analysis is collected from the manufacturing line. The base raw material is a steel alloy. Details such as water, electricity, oil and coolants, and their consumption details are also gathered. Welding and turning are the basic manufacturing processes. The total base raw material used, total power consumption and total oil consumption details are collected. LCA is done with the collected inventory details.

Stage 3: Impact assessment stage. The Eco-indicator 99 (I) method is used as the life cycle impact assessment (LCIA) method. Impact groups like carcinogens, respiratory organics, respiratory inorganics, climate change, ozone layer, ecotoxicity, radiation, acidification/eutrophication, land use and minerals are measured using this impact assessment method. The impact contribution of all factors for the welding process are shown in Figure 9.7.

Stage 4: Interpretation stage. In this stage, the results of the LCA are presented as well as the scope for further improvement. The overall environmental impacts for the welding process were computed as 56 Pt as shown in Figure 9.8. Based on the overall environmental impacts, it is possible to find the impacts for three categories: ecosystem quality, human health and resources.

9.4 Summary

This chapter illustrates three software modules for LCA and sustainable manufacturing. The first module is a product-based sustainability analysis module which highlights scope, impact parameters and procedural steps. The second and third modules are process-based modules (GaBi and SimaPro), wherein the procedural steps based on four phases of LCA are discussed with illustrations.

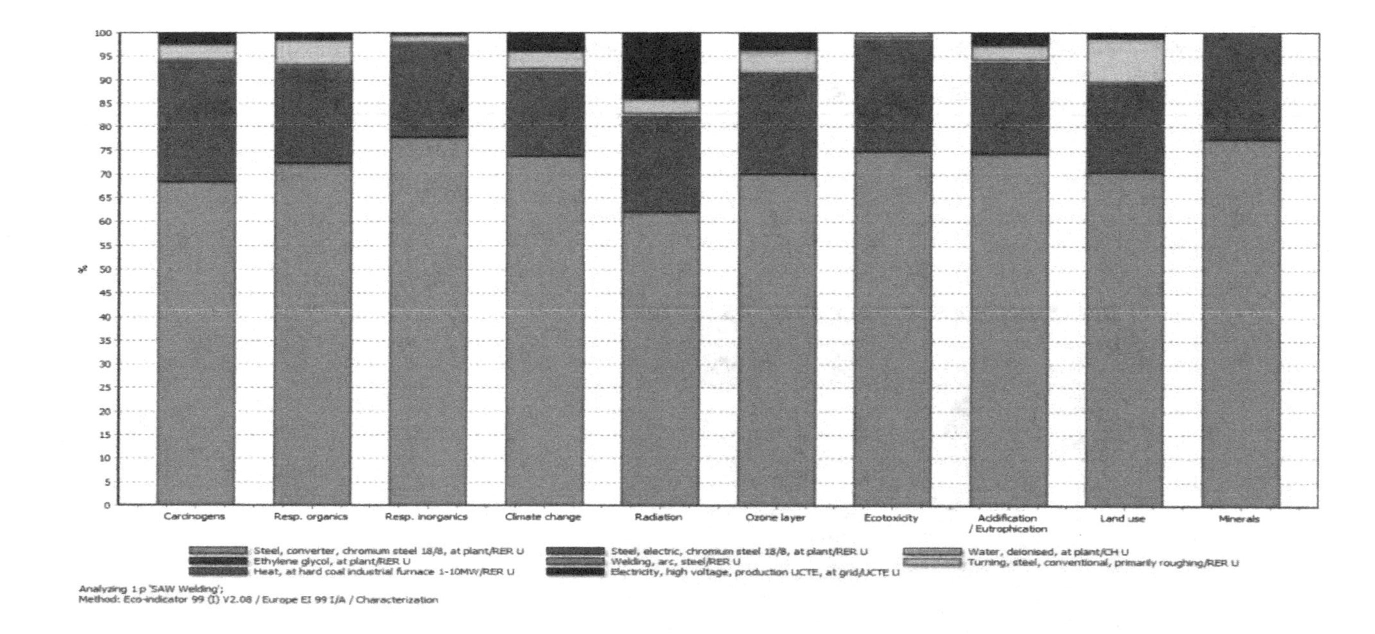

FIGURE 9.7
Impact contribution of all factors for welding process.

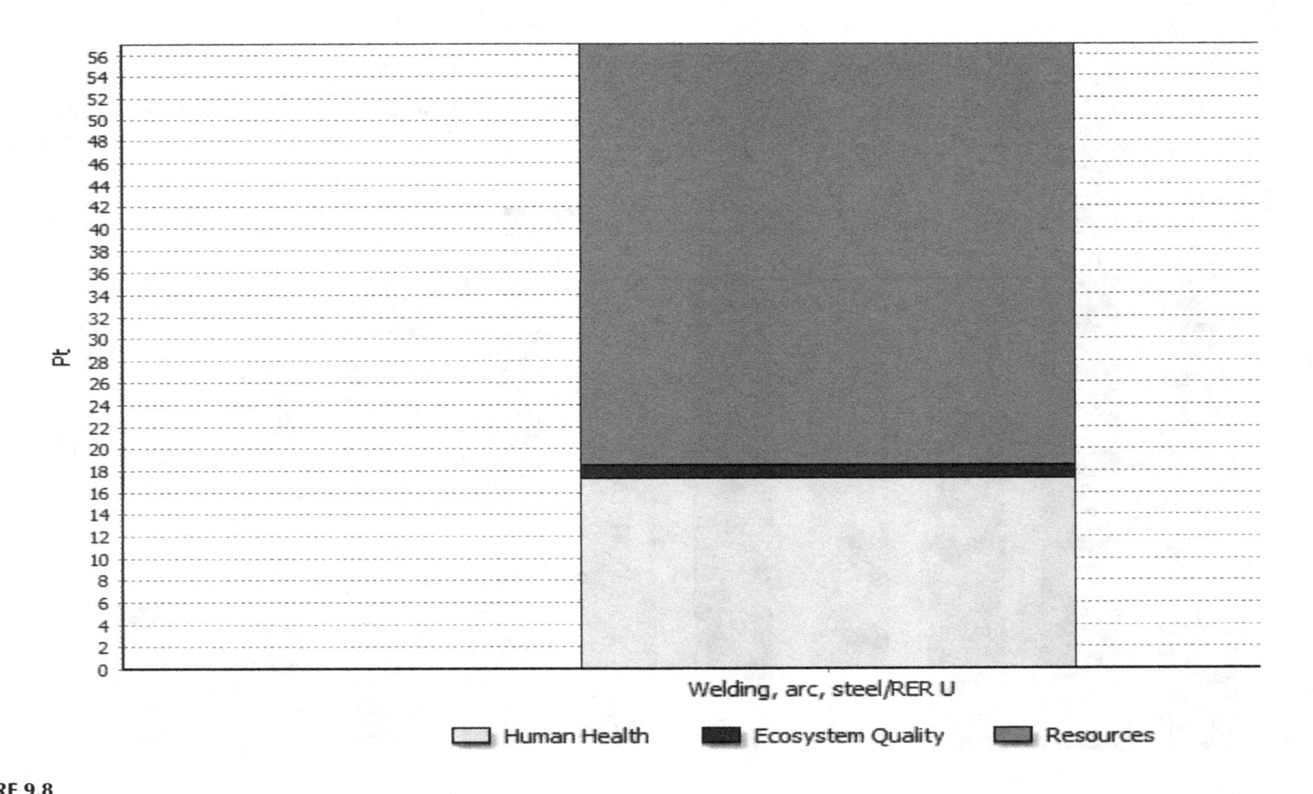

FIGURE 9.8
Environmental impact categories for welding process (single score).

References

Dassault Systèmes. (2010). *SolidWorks sustainability: An introduction to sustainable design*. Waltham, MA: Dassault Systèmes SolidWorks Corporation.

Goedkoop, M., Oele, M., Leijting, J., Ponsioen, T., & Meijer, E. (2016). Introduction to LCA with SimaPro title: Introduction to LCA with SimaPro introduction to LCA with SimaPro. Retrieved January 29, 2018, from: www. pre-sustainability. com.

Gruber, L.M., Brandstetter, C.P., Bos, U., Lindner, J.P., & Albrecht, S. (2016). LCA study of unconsumed food and the influence of consumer behavior. *The International Journal of Life Cycle Assessment, 21*(5), 773–784.

Martínez-Rocamora, A., Solís-Guzmán, J., & Marrero, M. (2016). LCA databases focused on construction materials: A review. *Renewable and Sustainable Energy Reviews, 58*, 565–573.

Morbidoni, A., Favi, C., & Germani, M. (2011). CAD-Integrated LCA tool: Comparison with dedicated LCA software and guidelines for the improvement. In Jürgen Hesselbach and Christoph Herrmann (Eds.), *Glocalized solutions for sustainability in manufacturing* (pp. 569–574). Berlin, Heidelberg: Springer.

Paudel, A.M., & Fraser, J.M. (2013). Teaching sustainability in an engineering graphics class with solid modeling tool. *Age, 23*, 1.

PE International. (2012). Retrieved from http://www.gabi-software.com.

PRé Consultants. (2008). SimaPro software. *SimaPro Version.*

Rodriguez, J., & Christensen, N. (2014). Modules to introduce sustainability and life cycle assessment (LCA) to undergraduate students. In *12th LACCEI Latin American and Caribbean conference engineers technology*, Guayaquil, Ecuador.

Spatari, S., Betz, M., Florin, H., Baitz, M., & Faltenbacher, M. (2001). Using GaBi 3 to perform life cycle assessment and life cycle engineering. *The International Journal of Life Cycle Assessment, 6*(2), 81.

Starostka-Patyk, M. (2015). New products design decision making support by SimaPro software on the base of defective products management. *Procedia Computer Science, 65*, 1066–1074.

10

Sustainability and Energy Aspects of Manufacturing Processes

Sustainability in manufacturing can be attained through material, product or process orientations with minimum impact on the environment. Major contributors to environmental impacts are material and energy consumption during machining operations. Among the three orientations, the process dimension is vital as it has a direct effect on resource consumption, emissions and energy utilization. This chapter presents sustainability and energy aspects of conventional, unconventional and additive manufacturing (AM) processes.

10.1 Sustainability of Conventional Manufacturing Processes

The details of certain research studies and inferences are presented in the following sections.

10.1.1 Framework-Based Studies

A comprehensive architecture for sustainability evaluation considering triple bottom line (TBL) dimensions was proposed with both qualitative and quantitative indicators such as energy consumption, CO_2 emissions, product satisfaction and health risks (Saad et al., 2019).

Bhanot et al. (2016) contributed a sustainability evaluation architecture for the turning process. The considered indicators included production cost, cutting quality, worker health and worker safety.

10.1.2 Energy-Based Studies

Emissions created by machine tools' energy consumption and cutting tools' embodied energy was examined in an experimental-type study. An on-line tool inspection system was used for measuring flank wear. A fluke power analyzer was used to quantify power consumption (Liu et al., 2016).

In empirical-based studies, equations can be used for calculation. Research by Kim and Dale (2008) considered the impact of non-renewable energy consumption, GHG emissions, acidification and eutrophication.

10.1.3 Life Cycle Assessment (LCA)-Based Studies

Operations considered – milling, laser cutting, cryogenic assisted machining.

Software modules used – GaBi, SimaPro.

Life cycle impact assessment (LCIA) method used – CML.

Database – ecoinvent.

Impact parameters – global warming potential (GWP), eutrophication potential (EP), acidification potential (AP) and so on.

10.1.4 Inferences from Certain Research Studies

- Pertaining to various approaches of milling, an average size mill produces higher impact than a micro mill, slow speeds produce higher impacts compared to high speeds and usage of a high-speed steel (HSS) tool produces a higher impact than a carbide tool due to slow speed criteria (de Souza Zanuto et al., 2019).
- Cryogenic cooling jet assisted turning was better than dry cutting with regard to eco-friendliness, clean environment and energy saving (Mia et al., 2019).

10.2 Sustainability of Unconventional Manufacturing Processes

Sustainable production focuses on efficiency optimization with minimization of environmental effects coupled with resource optimization (Rajurkar et al., 2017). There has been growing interest in sustainability of unconventional manufacturing processes (Gamage & DeSilva, 2015).

Among several unconventional manufacturing processes, the sustainability aspects of three processes – electrochemical machining (ECM), electric discharge machining (EDM) and ultrasonic machining (USM) – are described next (Rajurkar et al., 2017).

10.2.1 Sustainability Aspects of ECM, EDM and USM Processes

- ECM focuses on reducing waste generation to conserve energy. Electrolyte splashing and skin and eye irritation create health issues and impact operator safety. ECM creates sludge during machining with harmful contaminants.

- As the potential for hazard outcomes exist with EDM, various protective measures should be focused on the health and safety of operators. Dielectric properties are vital and their optimal levels and effect on sustainability need to be evaluated. Sustainability of dielectrics was assessed based on the following parameters: flash point, pour point, volatility, viscosity and so on. Water- and air-based dielectrics were eco-friendlier than oil based. EDM consumes more energy than traditional processes (turning) (Rajurkar et al., 2017).
- Sustainability issues in USM include prevalence of an electromagnetic field, ultrasonic frequencies and abrasive slurry. Experimental methods and mathematical modelling techniques were used to study sustainability of UCM (Gamage & Desilva, 2015).

10.2.2 LCA-Based Studies

LCA modules used – SimaPro, GaBi

Inventory database – ecoinvent

Impact parameters – GWP, ozone depletion potential (ODP), AP, EP, energy

Impact assessment methods used – Eco-indicator 99, ReCiPe, Co_2PE, USEPA TRACI, CML, IMPACT 2002+

Processes examined – EDM, electron beam machining (EBM), laser-based processes

Research studies referenced – Zhao et al. (2009), Zhao et al. (2010), Liu et al. (2018)

10.3 Sustainability of Additive Manufacturing Processes

Sustainability of additive manufacturing (AM) is focused on low carbon emissions, minimal energy consumption and other associated benefits (Gebler et al., 2014). Sustainable AM enables design freedom, optimized product geometries, simplified assemblies, enhanced resource efficiency, reduced energy consumption, lightweight products and improved operational efficiency (Mehrpouya et al., 2019).

10.3.1 LCA Studies of AM

AM processes studied – fused deposition modelling (FDM), laser engineered net shaping (LENS), stereolithography apparatus (SLA), powder bed fusion.

LCA modules used – GaBi, SimaPro.

LCIA methods used – ReCiPe, Eco-indicator

Inventory databases – ecoinvent, Emergy

Impact parameters – GWP, AP, ecotoxicity (ET)

Research studies referred to – Kreiger and Pearce (2013), Ma et al. (2018), Jiang et al. (2019), Faludi et al. (2019)

Inferences from studies:

- Distributed manufacturing based on open-source 3D printers had the scope with lesser environmental impact than traditional manufacturing for product variants (Kreiger & Pearce, 2013).
- The AM stage has the most influence on economic and environmental orientations of sustainability and the end-of-life (EoL) stage had more social impact (Ma et al., 2018).
- The LENS process was more sustainable than CNC machining for gear manufacture (Jiang et al., 2019).

10.3.2 Energy Evaluation in AM

Processes studied – selective laser sintering (SLS), FDM, binder jetting (BJ).

Energy parameters considered – process productivity; energy consumption rate; current; voltage; power; mean power consumption; maximum power; minimum power for warmup, build, and cooldown stages; electrical power consumption.

Measurement devices used – three single-phase clamp-on ammeters (for current measurement), LabVIEW circuit (for power measurement), NI-DAQ (data acquisition), Yokogawa CW240 power meter, Standby Energy-Monitor 'SEM 16+ USB'

Research studies referred to – Sreenivasan and Bourell (2010), Sreenivasan et al. (2010), Baumers et al. (2011), Junk and Cote (2012).

Inferences from studies:

- SLS holds potential from a sustainability perspective due to its lesser energy consumption.
- Energy consumption of fused layered manufacturing (FLM) technology was more than 3D printer technology.

10.3.3 Design for Additive Manufacturing (DFAM) Guidelines

The need and importance of design for additive manufacturing (DFAM) are detailed as follows (Thompson et al., 2016).

AM can develop various feature types and induce various constraints than other production processes. Hence, AM needs various process-specific design rules and tools.

AM processes have varied batch sizes, production times and cost drivers compared to conventional processes and hence a unique specialization is needed to facilitate DFAM.

The unique attributes of AM processes enable and need various methods for the design process and design practice. This encompasses new methods to find large, complex design spaces and to include material and multi-scale design concerns (Thompson et al., 2016).

As FDM is widely used in 3D printers, DFAM pertaining to FDM (Asadollahi-Yazdi et al., 2017) are presented in the following:

- Adapting part volume to machine capacity.
- Avoiding sharp edges and utilizing fillet to enhance product strength.
- Optimization of layer thickness and part orientation with effects on build time and roughness.
- It is desirable to provide a stereolithography file by smaller triangles to enable a source file with possible accuracy.

A design guideline is segmented into three parts: process characteristics, design principles and design rules (Leutenecker-Twelsiek et al., 2016).

- *Process characteristics:* They enumerate the fundamental knowledge on the working principle pertaining to a process for design and should be known by the designer for designing AM components.
- *Design principles*: They enable the designer to transform a principle solution into a specific, manufacturable design. They facilitate the designer to use AM's design freedom and to creatively avoid AM constraints.
- *Design rules:* The design rules address important characteristic facts and concepts for designers to design parts with good feasibility for AM.

Parts being produced (either existing or new) can utilize the advantages of AM by adopting two different design strategies: manufacturing driven or function driven. The details regarding the strategies could found in research studies by Klahn et al. (2015) and Vayre et al. (2012).

10.4 Summary

This chapter provides readers with insights on sustainability and energy efficiency of manufacturing processes. First, sustainability of conventional manufacturing is presented in three aspects: framework, energy and life cycle assessment (LCA)-based studies. Second, sustainability of unconventional manufacturing processes – ECM, EDM and USM – are presented. Third, sustainability of additive manufacturing processes is detailed from the viewpoint of LCA and energy evaluation. The scope of design for additive manufacturing (DFAM) and DFAM guidelines are discussed.

References

Asadollahi-Yazdi, E., Gardan, J., & Lafon, P. (2017). Integrated design for additive manufacturing based on skin-skeleton approach. *Procedia CIRP, 60*, 217–222.

Baumers, M., Tuck, C., Bourell, D.L., Sreenivasan, R., & Hague, R. (2011). Sustainability of additive manufacturing: Measuring the energy consumption of the laser sintering process. *Proceedings of the Institution of Mechanical Engineers, Part B: Journal of Engineering Manufacture, 225*(12), 2228–2239.

Bhanot, N., Rao, P.V., & Deshmukh, S.G. (2016). An assessment of sustainability for turning process in an automobile firm. *Procedia CIRP, 48*, 538–543.

de Souza Zanuto, R., Hassui, A., Lima, F., & Dornfeld, D.A. (2019). Environmental impacts-based milling process planning using a life cycle assessment tool. *Journal of Cleaner Production, 206*, 349–355.

Faludi, J., Van Sice, C.M., Shi, Y., Bower, J., & Brooks, O.M. (2019). Novel materials can radically improve whole-system environmental impacts of additive manufacturing. *Journal of Cleaner Production, 212*, 1580–1590.

Gamage, J.R., & DeSilva, A.K.M. (2015). Assessment of research needs for sustainability of unconventional machining processes. *Procedia CIRP, 26*, 385–390.

Gebler, M., Uiterkamp, A.J.S., & Visser, C. (2014). A global sustainability perspective on 3D printing technologies. *Energy Policy, 74*, 158–167.

Jiang, Q., Liu, Z., Li, T., Cong, W., & Zhang, H.C. (2019). Emergy-based life-cycle assessment (Em-LCA) for sustainability assessment: A case study of laser additive manufacturing versus CNC machining. *The International Journal of Advanced Manufacturing Technology, 102*(9–12), 4109–4120.

Junk, S., & Côté, S. (2012). A practical approach to comparing energy effectiveness of rapid prototyping technologies. In *Proceedings of the AEPR'12, 17th European forum on rapid prototyping and manufacturing*, Ecole Centrale, Paris.

Kim, S., & Dale, B.E. (2008). Life cycle assessment of fuel ethanol derived from corn grain via dry milling. *Bioresource Technology, 99*(12), 5250–5260.

Klahn, C., Leutenecker, B., & Meboldt, M. (2015). Design strategies for the process of additive manufacturing. *Procedia CIRP, 36*, 230–235.

Kreiger, M., & Pearce, J.M. (2013). Environmental life cycle analysis of distributed 3-D printing and conventional manufacturing of polymer products. *ACS Sustainable Chemistry and Engineering, 1*(12), 1511–1519.

Leutenecker-Twelsiek, B., Klahn, C., & Meboldt, M. (2016). Considering part orientation in design for additive manufacturing. *Procedia CIRP, 50*, 408–413.

Liu, Z.Y., Guo, Y.B., Sealy, M.P., & Liu, Z.Q. (2016). Energy consumption and process sustainability of hard milling with tool wear progression. *Journal of Materials Processing Technology, 229*, 305–312.

Liu, Z., Jiang, Q., Cong, W., Li, T., & Zhang, H.C. (2018). Comparative study for environmental performances of traditional manufacturing and directed energy deposition processes. *International Journal of Environmental Science and Technology, 15*(11), 2273–2282.

Ma, J., Harstvedt, J.D., Dunaway, D., Bian, L., & Jaradat, R. (2018). An exploratory investigation of additively manufactured product life cycle sustainability assessment. *Journal of Cleaner Production, 192*, 55–70.

Mehrpouya, M., Dehghanghadikolaei, A., Fotovvati, B., Vosooghnia, A., Emamian, S.S., & Gisario, A. (2019). The potential of additive manufacturing in the smart factory industrial 4.0: A review. *Applied Sciences, 9*(18), 3865.

Mia, M., Gupta, M.K., Lozano, J.A., Carou, D., Pimenov, D.Y., Królczyk, G., ... Dhar, N.R. (2019). Multi-objective optimization and life cycle assessment of eco-friendly cryogenic N2 assisted turning of Ti-6Al-4V. *Journal of Cleaner Production, 210*, 121–133.

Rajurkar, K.P., Hadidi, H., Pariti, J., & Reddy, G.C. (2017). Review of sustainability issues in non-traditional machining processes. *Procedia Manufacturing, 7*, 714–720.

Saad, M.H., Nazzal, M.A., & Darras, B.M. (2019). A general framework for sustainability assessment of manufacturing processes. *Ecological Indicators, 97*, 211–224.

Sreenivasan, R., & Bourell, D. (2010). Sustainability study in selective laser sintering- an energy perspective. *Minerals, Metals and Materials Society/AIME, 420 Commonwealth Dr., P. O. Box 430 Warrendale PA 15086 USA.[np]. 14–18 February.*

Sreenivasan, R., Goel, A., & Bourell, D.L. (2010). Sustainability issues in laser-based additive manufacturing. *Physics Procedia, 5*, 81–90.

Thompson, M.K., Moroni, G., Vaneker, T., Fadel, G., Campbell, R.I., Gibson, I., ... Martina, F. (2016). Design for additive manufacturing: Trends, opportunities, considerations, and constraints. *CIRP Annals, 65*(2), 737–760.

Vayre, B., Vignat, F., & Villeneuve, F. (2012). Designing for additive manufacturing. *Procedia CIRP, 3*, 632–637.

Zhao, F., Bernstein, W.Z., Naik, G., & Cheng, G.J. (2010). Environmental assessment of laser assisted manufacturing: Case studies on laser shock peening and laser assisted turning. *Journal of Cleaner Production, 18*(13), 1311–1319.

Zhao, F., Naik, G., & Zhang, L. (2009, January). Environmental sustainability of Laser-Based Manufacturing: Case studies on laser shock peening and laser assisted turning. In *ASME 2009 international manufacturing science and engineering conference* (pp. 97–105). American Society of Mechanical Engineers Digital Collection, West Lafayette, IN.

11

Case Studies on ECQFD, LCA and MCDM

11.1 Case on Environmentally Conscious Quality Function Deployment (ECQFD)

1. *Identification of voice of customer (VOC) and engineering metrics (EMs)*

   This step includes identifying customer requirements in environmental perspective so that they can be integrated in the design stage of the product life cycle. Then, those VOCs are converted into engineering metrics (EMs).

2. *Environmental VOCs*

   Environmental VOCs are the customer requirements which must be included in the product life cycle in the design phase itself to solve environment-related problems. These requirements include reliability of relays, ease of disposal and ease of cleaning of relays, long lifespan, safety, no sticking, lightweight, easy to install, not noisy, reduced energy consumption, and ease of operation.

3. *Environmental EMs*

   Environmental EMs includes technical properties of relays which need to be considered to ensure VOCs. They include maximum switching capacity, creeping distance, rated carry current, operating speed, operating force, wiping action, toxicity of material, process enhancement, biodegradability, reduced waste, reduced volume, recycling potential, insulation potential and serviceability.

4. *Identification of opportunities for design improvement*

   This includes phases I and II of ECQFD.

   Phase I

   In phase I, mapping needs to be done between VOCs and EMs. VOCs are written in rows and EMs are written in columns (Table 11.1). The relational strength needs to be assigned between a VOC and EM based on the scale given in Chapter 2, Section 2.1. This relational strength helps designers in decision making. The raw score and

TABLE 11.1

ECQFD Phase I

Engineering Metrics → / **Voice of Customer** ↓	**Customer weight**	**Maximum switching capacity**	**Opening force**	**Opening speed**	**Clearance**	**Creeping distance**	**Rated carry current**	**Wiping action**	**Serviceability**	**Toxicity of material**	**Biodegradability**	**Process enhancement**	**Lesser weight**	**Reduced volume**	**Recycling potential**	**Reduced waste**	**Insulation potential**
Reliability	9	3	3	3	—	—	3	3	3	—	—	3	9	9	3	—	9
No sticking	3	3	3	3	3	3	3	—	3	—	—	3	—	—	—	—	3
Moderate price	3	3	—	—	—	—	3	—	3	—	—	3	3	9	—	—	3
Compact/lightweight	9	3	3	3	–	–	3	3	3	—	—	3	9	9	3	3	—
Durable	3	3	3	3	3	3	9	3	3	—	—	3	9	9	—	—	—
Ease of disposal	3	—	—	—	—	—	—	—	3	—	—	3	3	3	9	9	—
Ease of cleaning	3	3	—	—	3	3	—	3	—	—	—	—	—	3	3	3	—
Easy to install	3	9	3	3	9	9	3	3	3	—	—	—	—	—	—	—	—
Long lifespan	9	9	3	3	3	—	9	3	3	3	3	9	3	3	3	3	9
Less noise	3	3	3	3	—	3	3	3	3	—	—	3	3	—	—	—	3
Reduced energy consumption	9	—	—	—	—	—	3	—	3	—	—	9	9	9	3	—	—
Ease of operation	3	3	9	9	3	3	3	3	3	—	—	9	3	3	—	—	3
Safety	3	3	3	3	—	—	9	3	—	—	—	3	—	—	—	—	3
Raw Score		225	153	153	90	72	261	135	171	27	27	297	333	351	144	90	207
Relative Weight		0.082	0.056	0.056	0.033	0.026	0.095	0.049	0.063	0.010	0.010	**0.109**	**0.122**	**0.128**	0.053	0.033	0.076

relative score of each EM are then evaluated using the equation given in the methodology section (Chapter 2). A sample calculation for an EM (maximum switching capacity) is given next:

Raw score of EM_1 (Maximum Switching Capacity)

Raw score of EM_1 = [(9*3) + (3*3) + (3*3) + (9*3) + (3*3) + (3*0) + (3*3) + (3*9) + (9*9) + (3*3) + (9*0) + (3*3) + (3*3)]

= (27 + 9 + 9 + 27 + 9 + 0 + 9 + 27 + 81+ 9 + 0 + 9 + 9)

Raw score of $EM_1 = 225$

Relative weight of EM_1 (Maximum Switching Capacity)

$$\text{Relative weight of } EM_1 = \frac{225}{2736} = 0.082$$

Phase II

Phase II includes the integration of EMs in product components. In phase II, the EMs are written in rows and product components are written in columns (Table 11.2). The relative weight of each EM is taken from phase I. The relative weight of each component is

TABLE 11.2

ECQFD Phase II

Engineering Metrics ↓	Phase I Relative Weight	Components			
		Frame	Coil	Armature	Contacts
Maximum switching capacity	0.0822	3	3	3	9
Opening force	0.0559	9	3	3	3
Opening speed	0.0559	9	3	3	9
Clearance	0.0329	3	3	3	3
Creeping distance	0.0263	9	3	3	9
Rated carry current	0.0954	3	3	9	9
Wiping action	0.0493	3	9	9	3
Serviceability	0.0625	3	3	3	9
Toxicity of material	0.0099	3	3	3	9
Biodegradability	0.0099	3	—	—	3
Process enhancement	0.1086	9	3	3	9
Lesser weight	0.1217	9	3	9	9
Reduced volume	0.1283	9	3	9	9
Recycling potential	0.0526	3	3	3	9
Reduced waste	0.0329	3	3	3	3
Insulation potential	0.0757	3	3	3	9
Raw Score		5.9802	3.2661	5.3385	7.9146
Relative Weight		**0.265794**	0.145164	**0.237273**	**0.351769**

calculated the same as in phase I. A sample calculation for a component (Frame) is shown next:

Raw score of Component (Frame)

Raw score of C_1 = {(0.0822*3) + (0.0559*9) + (0.0559*9) + (0.0329*3) + (0.0263*9) + (0.0954*3) + (0.0493*3) + (0.0625*3) + (0.0099*3) + (0.0099*3) + (0.1086*9) + (0.1217*9) + (0.1283*9) + (0.0526*3) + (0.0329*3) + (0.0757*3)}

Raw score of C_1 = 5.9802

Relative weight of Component (Frame)

$$\text{Relative weight of } C_1 = \frac{5.9802}{22.5} = 0.265$$

The top three engineering metrics are reduced volume, lesser weight and process enhancement. The top three components are contacts, frame and armature.

Options for ECQFD

Option I

- The material of the frame should be such that it can support the parts of the relay without any failure.
- The material of the coil wire should be such that it can withstand a heavy load.
- The armature is the moving part in the relay so it should be made lightweight and using high strength material.

Option II

- The material of the contacts should have high conductivity and should be non-sticking to avoid opening and closing of the circuit.
- The lifetime of the armature should be long.
- The movement of armature should be minimum.

5. *Evaluation of considered design improvements*

This includes phases III and IV of ECQFD.

Phase III

In this phase, for developed options, the impact of design improvement on EMs needs to be evaluated. In this phase, the designer can make several alternative options for improvement, and then the improvement rate of EMs are evaluated using the equation given in the methodology section (Chapter 2, Section 2.3). Two options have been considered in this study and the improvement rate of EMs for each option have been calculated (Tables 11.3 and 11.4). The sample calculation for an EM (Maximum Switching Capacity) is shown next:

TABLE 11.3

ECQFD Phase III for Option I

Engineering Metrics ↓	Phase I Relative Weight	Components				Score	Improvement Rate of EM
		Frame	Coil	Armature	Contacts		
Maximum switching capacity	0.0822	3	3			6	0.33
Opening force	0.0559	9	3			12	0.67
Opening speed	0.0559					0	0
Clearance	0.0329					0	0
Creeping distance	0.0263					0	0
Rated carry current	0.0954					0	0
Wiping action	0.0493					0	0
Serviceability	0.0625			3		3	0.17
Toxicity of material	0.0099	3				3	0.17
Biodegradability	0.0099					0	0
Process enhancement	0.1086					0	0
Lesser weight	0.1217			9		9	0.3
Reduced volume	0.1283			9		9	0.3
Recycling potential	0.0526					0	0
Reduced waste	0.0329					0	0
Insulation potential	0.0757					0	0

TABLE 11.4
ECQFD Phase III for Option II

Engineering Metrics ↓	Phase I Relative Weight	Components				Score	Improvement Rate of EM
		Frame	Coil	Armature	Contacts		
Maximum switching capacity	0.0822				9	9	0.5
Opening force	0.0559					0	0
Opening speed	0.0559					0	0
Clearance	0.0329				3	3	0.25
Creeping distance	0.0263			3		3	0.125
Rated carry current	0.0954					0	0
Wiping action	0.0493			9		9	0.38
Serviceability	0.0625			3		3	0.17
Toxicity of material	0.0099					0	0
Biodegradability	0.0099					0	0
Process enhancement	0.1086					0	0
Lesser weight	0.1217			9		9	0.3
Reduced volume	0.1283			9		9	0.3
Recycling potential	0.0526					0	0
Reduced waste	0.0329					0	0
Insulation potential	0.0757				9	9	0.5

Improvement rate of engineering metric – Maximum Switching Capacity (IR_1)

$$(IR_1) = \frac{(3*1)+(3*1)+(3*0)+(9*0)}{3+3+3+9} = 0.333$$

Phase IV

The matrix structure of this phase is similar to that of phase I (Tables 11.5 and 11.6). This phase involves calculation of the improvement rate of VOC based on the calculated improvement rate of EM. The equation for calculating the improvement rate of VOC is mentioned in the methodology section (Chapter 2, Section 2.4). A sample calculation for the improvement rate of customer requirements (Reliability) for option I and the global score is shown next:

The improvement rate of CR Reliability (IR_1)

$$IR_1 = \left(\frac{\begin{matrix}(3*0.333)+(3*0.667)+(3*0)+(0*0)+(0*0)+(3*0)+(3*0)+(3*0.167)+(0*0.167)+\\(0*0)+(3*0)+(9*0.3)+(9*0.3)+(3*0)+(0*0)+(9*0)\end{matrix}}{(3+3+3+0+0+3+3+3+0+0+3+9+9+3+0+9)*9} \right)$$

$$IR_1 = 0.0194$$

The improvement effect of customer requirement is calculated by multiplying the improvement rate of customer requirement (IR_i) with the customer weight W_i.

$$IE_i = IR_i * W_i$$

The improvement effect of CR Reliability (IE_1)

$$IE_1 = 0.0194 * 9 = 0.1745$$

The global score is calculated as the sum of the improvement effects of all customer requirements.

Global score = (0.1745 + 0.1296 + 0.1889 + 0.1978 + 0.1745 + 0.0767 + 0.0905 + 0.1310 + 0.1130 + 1467 + 0.1639 + 0.1722 + 0.1111)

Global score = 1.8703

Based on computations in ECQFD, the global score for option II is found to be higher. Based on this analysis, an alternative material for contacts is explored with high conductivity and non-sticking. Also, material for the armature is analyzed to be durable with a longer lifetime.

TABLE 11.5

ECQFD Phase IV for Option I

Engineering Metrics → Voice of Customer ↓	Customer weight	Maximum switching capacity	Opening force	Opening speed	Clearance	Creeping distance	Rated carry current	Wiping action	Serviceability	Toxicity of material	Biodegradability	Process enhancement	Lesser weight	Reduced volume	Recycling potential	Reduced waste	Insulation potential	Improvement rate of customer requirement	Improvement effect of customer requirement
Reliability	9	3	3	3	0	0	3	3	3	0	0	3	9	9	3	0	9	0.0194	0.1745
No sticking	3	3	3	3	3	3	3	0	3	0	0	3	0	0	0	0	3	0.0432	0.1296
Moderate price	3	3	0	0	0	0	3	0	3	0	0	3	3	9	0	0	3	0.0630	0.1889
Compact/lightweight	9	3	3	3	0	0	3	3	3	0	0	3	9	9	3	3	0	0.0220	0.1978
Durable	3	3	3	3	3	3	9	3	3	0	0	3	9	9	0	0	0	0.0582	0.1745
Ease of disposal	3	0	0	0	0	0	0	0	3	0	0	3	3	3	9	9	0	0.0256	0.0767
Ease of cleaning	3	3	0	0	3	3	0	3	0	0	0	0	0	3	3	3	0	0.0302	0.0905
Easy to install	3	9	3	3	9	9	3	3	3	0	0	0	0	0	0	0	0	0.0437	0.1310
Long lifespan	9	9	3	3	3	0	9	3	3	3	3	9	3	3	3	3	9	0.0126	0.1130
Less noise	3	3	3	3	0	3	3	3	3	0	0	3	3	0	0	0	3	0.0489	0.1467
Reduced energy consumption	9	0	0	0	0	0	3	0	3	0	0	9	9	9	3	0	0	0.0182	0.1639
Ease of operation	3	3	9	9	3	3	3	3	3	0	0	9	3	3	0	0	3	0.0574	0.1722
Safety	3	3	3	3	0	0	9	3	0	0	0	3	0	0	0	0	3	0.0370	0.1111
Improvement rate of engineering metrics		0.333	0.667	0	0	0	0	0	0.167	0.167	0	0	0.3	0.3	0	0	0		
Global Score																			1.8703

TABLE 11.6
ECQFD Phase IV for Option II

Engineering Metrics → **Voice of Customer ↓**	**Customer weight**	**Maximum switching capacity**	**Opening force**	**Opening speed**	**Clearance**	**Creeping distance**	**Rated carry current**	**Wiping action**	**Serviceability**	**Toxicity of material**	**Biodegradability**	**Process enhancement**	**Lesser weight**	**Reduced volume**	**Recycling potential**	**Reduced waste**	**Insulation potential**	**Improvement rate of customer requirement**	**Improvement effect of customer requirement**
Reliability	9	3	3	3	0	0	3	3	3	0	0	3	9	9	3	0	9	0.0284	0.2554
No sticking	3	3	3	3	3	3	3	0	3	0	0	3	0	0	0	0	3	0.0571	0.1713
Moderate price	3	3	0	0	0	0	3	0	3	0	0	3	3	9	0	0	3	0.0877	0.2630
Compact/lightweight	9	3	3	3	0	0	3	3	3	0	0	3	9	9	3	3	0	0.0210	0.1894
Durable	3	3	3	3	3	3	9	3	3	0	0	3	9	9	0	0	0	0.0631	0.1892
Ease of disposal	3	0	0	0	0	0	0	0	3	0	0	3	3	3	9	9	0	0.0256	0.0767
Ease of cleaning	3	3	0	0	3	3	0	3	0	0	0	0	0	3	3	3	0	0.0738	0.2214
Easy to install	3	9	3	3	9	9	3	3	3	0	0	0	0	0	0	0	0	0.0754	0.2262
Long lifespan	9	9	3	3	3	0	9	3	3	3	3	9	3	3	3	3	9	0.0212	0.1909
Less noise	3	3	3	3	0	3	3	3	3	0	0	3	3	0	0	0	3	0.0656	0.1967
Reduced energy consumption	9	0	0	0	0	0	3	0	3	0	0	9	9	9	3	0	0	0.0182	0.1639
Ease of operation	3	3	9	9	3	3	3	3	3	0	0	9	3	3	0	0	3	0.0466	0.1398
Safety	3	3	3	3	0	0	9	3	0	0	0	3	0	0	0	0	3	0.0509	0.1528
Improvement rate of engineering metrics		0.5	0	0	0.25	0.125	0	0.375	0.167	0	0	0	0.3	0.3	0	0	0.5		
Global Score																			2.4366

11.2 Case on Life Cycle Costing and Life Cycle Assessment

This section presents two cases: one for life cycle costing and another for life cycle assessment (LCA).

11.2.1 Product Life Cycle Costing

The details of the life cycle costing (LCC) of an electronics product is shown in Table 11.7.

From Figure 11.1, part A and part C have the highest positive net product gain (ΔPG) and hence possess good potential for reverse processing.

11.2.2 Case Study on LCA

Stage 1: Goal and scope definition

The goal of the presented study is to perform LCA of helical gear. The scope includes the boundary condition. In this study, the focus is on the processing stage of the gear.

Stage 2: Inventory analysis stage

Input material: ABS P430

Input quantity: 38.88 gm

The stage of fabrication of the helical gear through fused deposition modelling (FDM) is as follows:

FDM printing followed by post-processing (cleaning).

Output from FDM – unfinished helical gear (38.88 gm).

Output from cleaning process – finished helical gear (36.17 gm).

Electricity used – 7250 kJ.

Operation on PC – 4 hours.

Stage 3: Impact assessment stage

Here, the Eco-indicator 99 method is considered for analysis.

Stage 4: Interpretation stage

In this stage, the results of LCA and the scope for further improvement are presented.

TABLE 11.7

Life Cycle Costing

	Part A	Part B	Part C	Part D	Part E	Part F	Part G	Part H	Part I	Part J
Producing a New Part										
Product cost	12	6	23	9	8	6	0.5	0.5	0.2	0.4
Environmental cost	1.2	0.6	2.3	0.9	0.8	0.6	0.05	0.05	0.02	0.04
Product life cycle cost (PLCC)	13.2	6.6	25.3	9.9	8.8	6.6	0.55	0.55	0.22	0.44
Market price (MP)	14	8	26	12	9	8	0.8	0.8	0.4	0.6
Product effectiveness (PE)	1	0	1	1	1	1	1	1	1	1
Product value (PV)	14	0	26	12	9	8	0.8	0.8	0.4	0.6
Product gain (PG_{NC})	0.8	–6.6	0.7	2.1	0.2	1.4	0.25	0.25	0.18	0.16
Remanufacturing an Old Part										
Procurement cost	4	2	7	3	2.3	2	0.15	0.15	0.1	0.15
Remanufacturing cost	5	3	8	4	3.5	2.5	0.175	0.175	0.15	0.18
Environmental cost	0.05	0.03	0.08	0.04	0.035	0.025	0.00175	0.00175	0.0015	0.0018
Product life cycle cost (PLCC)	9.05	5.03	15.08	7.04	5.835	4.525	0.32675	0.32675	0.2515	0.3318
Market price (MP)	14	8	26	12	9	8	0.8	0.8	0.4	0.6
Product effectiveness (PE)	1	0	1	1	1	1	1	1	1	1
Product value (PV)	14	0	26	12	9	8	0.8	0.8	0.4	0.6
Product gain (PG_{OC})	4.95	–5.03	10.92	4.96	3.165	3.475	0.47325	0.47325	0.1485	0.2682
$\Delta PG = PG_{OC} - PG_{NC}$	4.15	1.57	10.22	2.86	2.965	2.075	0.22325	0.22325	– 0.0315	0.1082

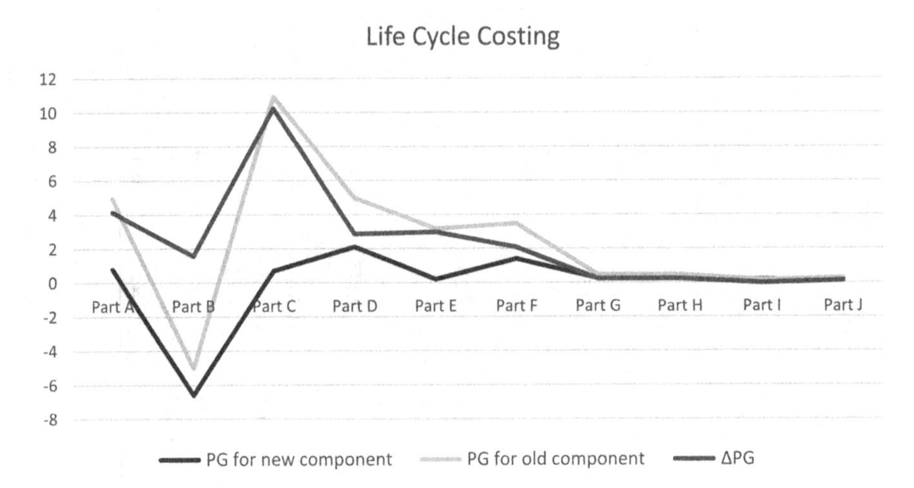

FIGURE 11.1
Product gain and net product gain values for different life cycle phases.

The flow chart for fabricating helical gear through FDM with environmental impact is presented in Figure 11.2. Figure 11.2 shows the total environmental impact generated in fabricating the helical gear is 99.7 mpt. Out of which environmental impact due to the electricity grid (73.1 mpt), computer use (19.2 mpt), processing (2.07 mpt) and ABS material (5.32 mpt) are noted.

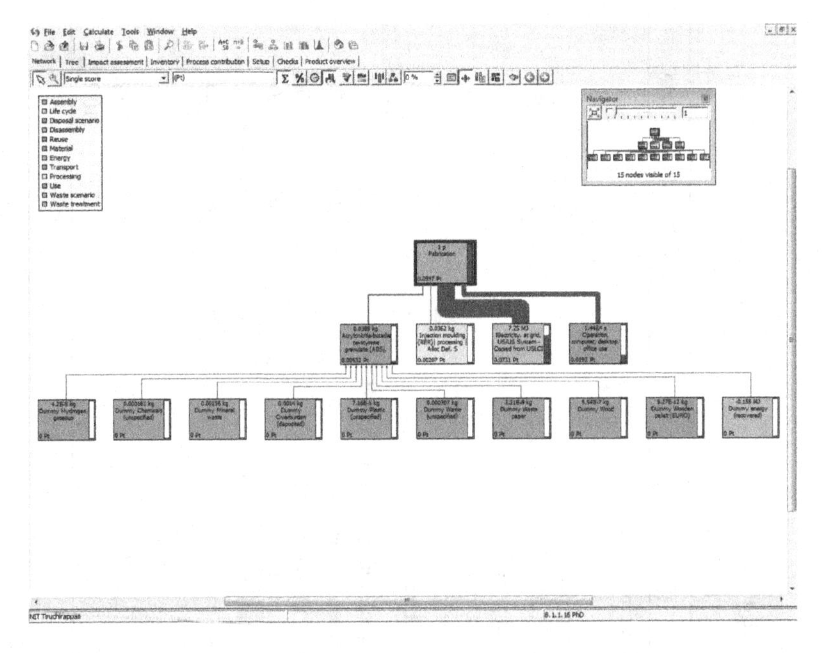

FIGURE 11.2
Flow chart for fabricating helical gear through FDM with environmental impact.

The environmental characterization for the case product in ten impact categories is shown in Figure 11.3. The red colour indicates environmental impact generated due to ABS material, green colour presents impact due to process, yellow colour indicates impact generated due to electricity and blue colour indicates impact generated due to computer use.

Figure 11.4 presents the environmental impact of the case product in three damage categories, namely human health, ecosystem quality and resources. The environmental impact on human health is found to be 87 mpt, that on ecosystem quality is 2.55 mpt and for resources it is found to be 10.1 mpt. It can be seen that environmental impact generated by fabrication of helical gear is affecting human health the most, followed by resources and ecosystem quality (Figure 11.5).

11.3 Case on MCDM

In this case, project selection for sustainable manufacturing (SM) has been done. Three potential projects are considered, namely LCA, design for

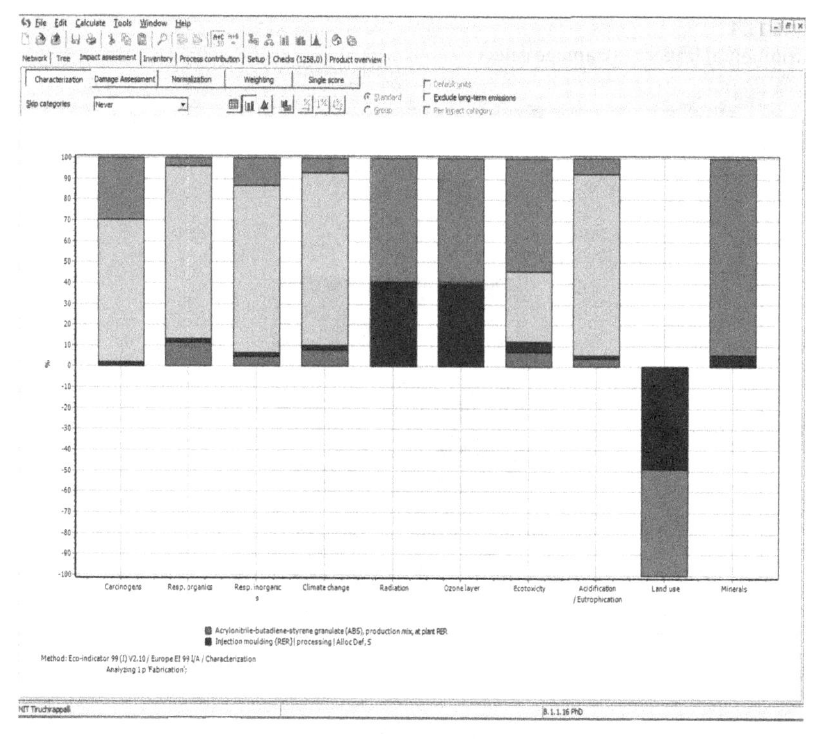

FIGURE 11.3
Characterization results in ten impact categories.

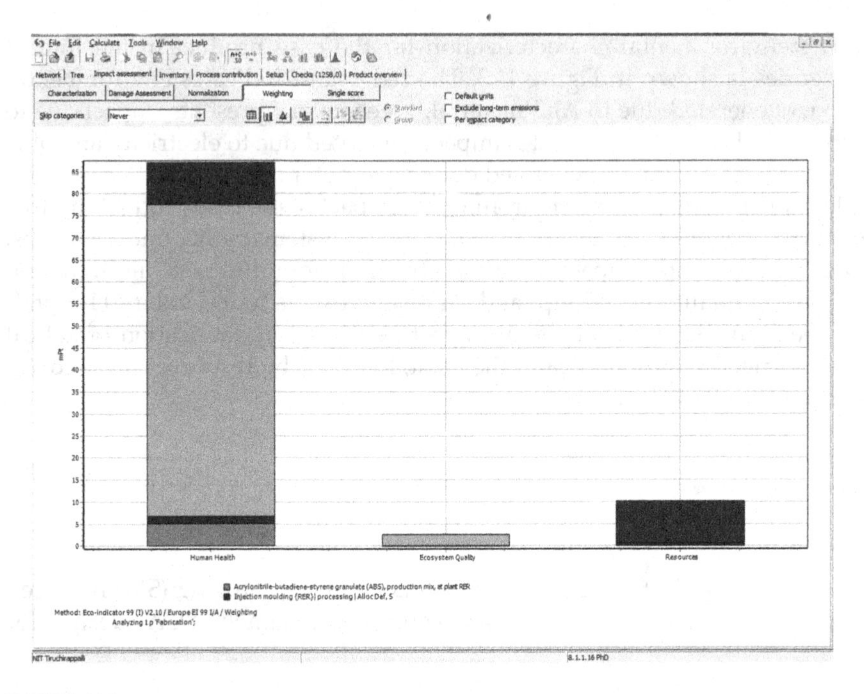

FIGURE 11.4
Environmental impact in damage categories.

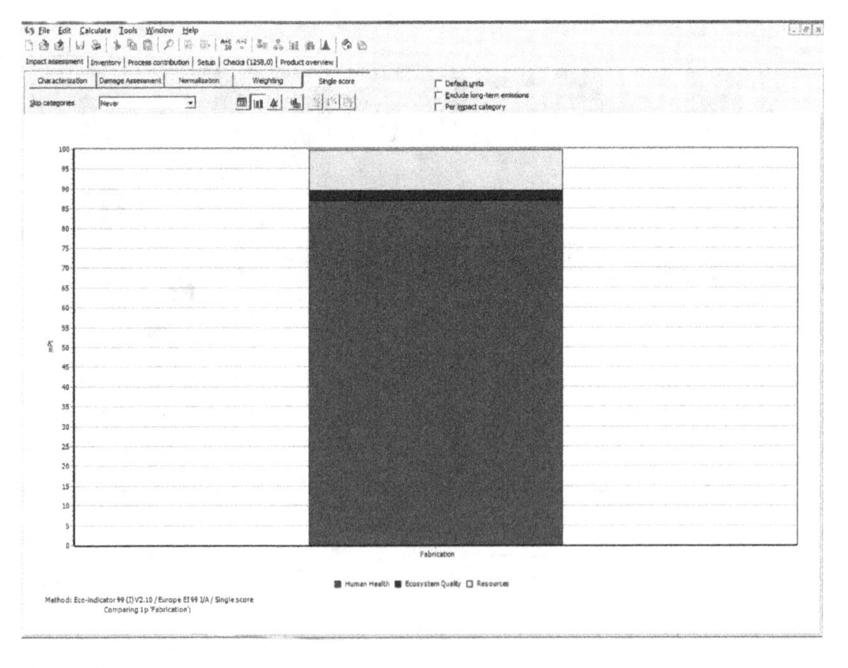

FIGURE 11.5
Environmental impact in single score.

environment (DFE) guidelines and ECQFD. To analyze the identified projects and to prioritize the projects, an analytic hierarchy process (AHP)-based multi-criteria decision making (MCDM) technique has been used. Twelve criteria have been applied for prioritizing SM projects. The criteria are health benefits, safety, skills enhancement, increase in employment opportunity, employment rate, implementation cost, profit, wastage, energy usage, water pollution, disposal action and air pollution. A pairwise comparison matrix has been made between criteria by collecting expert opinion. The pairwise comparison matrix is depicted in Table 11.8. The weight of each criterion is computed using AHP methodology. The weight of each criterion is presented in Table 11.9.

TABLE 11.8

Pairwise Comparison Matrix

	Health benefits	**Safety**	**Skills enhancement**	**Increase in employment opportunity**	**Employment rate**	**Implementation cost**	**Profit**	**Wastage**	**Energy usage**	**Water pollution**	**Disposal action**	**Air pollution**
Health benefits	1	1	1	1	1	1	1/3	3	1/3	3	1/3	3
Safety	1	1	1/3	1/3	1/3	3	1/3	3	1/3	3	1/3	3
Skills enhancement	1	3	1	1	1/3	3	3	3	1/3	3	1/3	3
Increase in employment opportunity	1	3	1	1	1	3	3	3	1/3	3	1	3
Employment rate	1	3	3	1	1	3	1	3	1/3	3	1/3	3
Implementation cost	1	1/3	1/3	1/3	1/3	1	1/3	1	1/3	1/3	1/3	3
Profit	3	3	1/3	1/3	1	3	1	3	3	3	1/3	3
Wastage	1/3	1/3	1/3	1/3	1/3	1	1/3	1	1/3	1	1/3	1
Energy usage	3	3	3	3	3	3	1/3	3	1	3	1	3
Water pollution	1/3	1/3	1/3	1/3	1/3	3	1/3	1	1/3	1	1/3	1
Disposal action	3	3	3	1	3	3	3	3	1	3	1	3
Air pollution	1/3	1/3	1/3	1/3	1/3	1/3	1/3	1	1/3	1	1/3	1

TABLE 11.9
Weightage of Criteria

	Health Benefits	Safety	Skills enhancement	Increase in employment opportunity	Employment rate	Implementation cost	Profit	Wastage	Energy usage	Water pollution	Disposal action	Air pollution	Weight
Health benefits	1	1	1	1	1	1	1/3	3	1/3	3	1/3	3	0.07
Safety	1	1	1/3	1/3	1/3	3	1/3	3	1/3	3	1/3	3	0.062
Skills enhancement	1	3	1	1	1/3	3	3	3	1/3	3	1/3	3	0.096
Increase in employment opportunity	1	3	1	1	1	3	3	3	1/3	3	1	3	0.11
Employment rate	1	3	3	1	1	3	1	3	1/3	3	1/3	3	0.1
Implementation cost	1	1/3	1/3	1/3	1/3	1	1/3	1	1/3	1/3	1/3	3	0.039
Profit	3	3	1/3	1/3	1	3	1	3	3	3	1/3	3	0.117
Wastage	1/3	1/3	1/3	1/3	1/3	1	1/3	1	1/3	1	1/3	1	0.032
Energy usage	3	3	3	3	3	3	1/3	3	1	3	1	3	0.153
Water pollution	1/3	1/3	1/3	1/3	1/3	3	1/3	1	1/3	1	1/3	1	0.038
Disposal action	3	3	3	1	3	3	3	3	1	3	1	3	0.153
Air pollution	1/3	1/3	1/3	1/3	1/3	1/3	1/3	1	1/3	1	1/3	1	0.03

Consistency check

$\lambda max = 13.53$

$CI = (\lambda max - N)/(N - 1)$

$CI = (13.53 - 12)/(12 - 1)$

$CI = 1.53/11 = 0.139$

$RI = 1.536$ (for $n = 12$)

$CR = CI/RI$

$CR = 0.139/1.536$

$CR = 0.090$

After calculating the weight of each criterion, the next step is to compare alternative projects with reference to each criterion. Tables 11.10–11.21 show the comparison of three alternatives based on each criterion.

TABLE 11.10

Weights of Alternatives Based on Criterion 'Health Benefits'

Health Benefits	LCA	DFE Guidelines	ECQFD	Weight
LCA	1	3	3	0.574
DFE Guidelines	1/3	1	3	0.286
ECQFD	1/3	1/3	1	0.14

Consistency check

λmax = 3.137

CI = 0.068

RI = 0.524 (for n = 12)

CR = 0.130

TABLE 11.11

Weights of Alternatives Based on Criterion 'Safety'

Safety	LCA	DFE Guidelines	ECQFD	Weight
LCA	1	3	3	0.574
DFE Guidelines	1/3	1	3	0.286
ECQFD	1/3	1/3	1	0.14

λmax = 3.137

CI = 0.068

CR = 0.130

TABLE 11.12

Weights of Alternatives Based on Criterion 'Skills Enhancement'

Skills Enhancement	LCA	DFE Guidelines	ECQFD	Weight
LCA	1	1	3	0.454
DFE Guidelines	1	1	1	0.321
ECQFD	1/3	1	1	0.225

Max = 3.136

CI = 0.068

CR = 0.130

TABLE 11.13

Weights of Alternatives Based on Criterion 'Increase in Employment Opportunity'

Increase in Employment Opportunity	LCA	DFE Guidelines	ECQFD	Weight
LCA	1	1	3	0.429
DFE Guidelines	1	1	3	0.429
ECQFD	1/3	1/3	1	0.143

$\lambda max = 3$

$CI = 0$

$CR = 0$

TABLE 11.14

Weights of Alternatives Based on Criterion 'Employment Rate'

Employment Rate	LCA	DFE Guidelines	ECQFD	Weight
LCA	1	3	3	0.6
DFE Guidelines	1/3	1	1	0.2
ECQFD	1/3	1	1	0.2

$\lambda max = 3$

$CI = 0$

$CR = 0$

TABLE 11.15

Weights of Alternatives Based on Criterion 'Implementation Cost'

Implementation Cost	LCA	DFE Guidelines	ECQFD	Weight
LCA	1	3	3	0.6
DFE Guidelines	1/3	1	1	0.2
ECQFD	1/3	1	1	0.2

$\lambda max = 3$

$CI = 0$

$CR = 0$

TABLE 11.16

Weights of Alternatives Based on Criterion 'Profit'

Profit	LCA	DFE Guidelines	ECQFD	Weight
LCA	1	1	3	0.454
DFE Guidelines	1	1	1	0.321
ECQFD	1/3	1	1	0.225

$\lambda max = 3.136$

CI = 0.068

CR = 0.130

TABLE 11.17

Weights of Alternatives Based on Criterion 'Wastage'

Wastage	LCA	DFE Guidelines	ECQFD	Weight
LCA	1	3	3	0.574
DFE Guidelines	1/3	1	3	0.286
ECQFD	1/3	1/3	1	0.14

$\lambda max = 3.137$

CI = 0.068

CR = 0.130

TABLE 11.18

Weights of Alternatives Based on Criterion 'Energy Usage'

Energy Usage	LCA	DFE Guidelines	ECQFD	Weight
LCA	1	3	3	0.574
DFE Guidelines	1/3	1	3	0.286
ECQFD	1/3	1/3	1	0.14

$\Lambda max = 3.137$

CI = 0.068

CR = 0.130

TABLE 11.19

Weights of Alternatives Based on Criterion 'Water Pollution'

Water Pollution	LCA	DFE Guidelines	ECQFD	Weight
LCA	1	3	3	0.6
DFE Guidelines	1/3	1	1	0.2
ECQFD	1/3	1	1	0.2

$\lambda max = 3$

CI = 0

CR = 0

TABLE 11.20

Weights of Alternatives Based on Criterion 'Disposal Action'

Disposal Action	LCA	DFE Guidelines	ECQFD	Weight
LCA	1	3	3	0.574
DFE Guidelines	1/3	1	3	0.286
ECQFD	1/3	1/3	1	0.14

$\lambda max = 3.137$

CI = 0.068

CR = 0.130

TABLE 11.21

Weights of Alternatives Based on Criterion 'Air Pollution'

Air Pollution	LCA	DFE Guidelines	ECQFD	Weight
LCA	1	3	3	0.574
DFE Guidelines	1/3	1	3	0.286
ECQFD	1/3	1/3	1	0.14

$\lambda max = 3.137$

CI = 0.068

CR = 0.130

The next step is to form a global matrix by combining the project weights for the criteria obtained in Table 11.20 through Table 11.22.

The ranking of projects can be done by multiplying the global matrix with the weights of the criteria. The obtained global weight for each project and prioritization of projects are depicted in Table 11.23.

Based on the priority order, the projects are subjected to deployment. LCA has been done to quantify environmental impacts associated with the candidate product. This is followed by deployment of DFE guidelines to ensure product designs with minimal environmental impact. Then, ECQFD is done to handle both traditional and environmental aspects effectively.

TABLE 11.22

Global Matrix

Project	C1	C2	C3	C4	C5	C6	C7	C8	C9	C10	C11	C12
P1	0.57	0.57	0.45	0.43	0.6	0.6	0.45	0.57	0.57	0.6	0.57	0.57
P2	0.28	0.28	0.32	0.43	0.2	0.2	0.32	0.28	0.28	0.2	0.28	0.28
P3	0.14	0.14	0.22	0.14	0.2	0.2	0.22	0.14	0.14	0.2	0.14	0.14

TABLE 11.23
Global Weight and Ranking

Project	Global Weight	Rank
LCA	0.537	1
DFE Guidelines	0.294	2
ECQFD	0.169	3

11.4 Summary

This chapter illustrates three case studies: first, on environmentally conscious quality function deployment (ECQFD) for an electronics product with details of four phases; second, on life cycle costing with the comparison of manufacturing scenarios and life cycle assessment of a case product, wherein the SimaPro module is used for performing LCA in four phases; and third, on multi-criteria decision making for project selection in sustainable manufacturing with the derivation of priority of projects.

12

Research Issues in Sustainable Manufacturing

Research issues and agenda in sustainable manufacturing (SM) are presented from five perspectives.

12.1 Relation between Lean and Sustainable Manufacturing

Lean initiatives facilitate sustainable manufacturing as waste elimination goal of lean enable optimized use of resources. When resource utilization is optimal, environmental impacts are minimal and ecological balance is ensured. As the ninth waste in lean pertains to environmental waste, pollutants and emissions need to be minimized. Lean concepts can be integrated with radio frequency identification (RFID) and sensor-based mechanisms to track inventory and time parameters in the value stream (Varela et al., 2019). Streamlined and synchronized processes enable waste elimination and value addition from the customer perspective. Lean tools such as 5S are extended to 7S to include safety and sustainability aspects. Value stream mapping (VSM), a popular lean tool, is extended as sustainable VSM to deal with water, energy and consumption of other resources to optimize processes to minimize the environmental impact. Also, VSM can be integrated with life cycle assessment (LCA) for comprehensive monitoring of resources and impacts (Vinodh et al., 2016).

12.2 Multi-Criteria Decision Making (MCDM) and Easily Implemented MCDM methods in Sustainable Manufacturing

MCDM methods have widespread application in SM from the viewpoint of alternative selection in terms of materials, product design, process selection, supplier selection, technology selection and so on. Various classes of MCDM methods are applied such as network-based, outranking, compromise solution, fuzzy-based, grey theory-based and hybrid techniques. A list of MCDM methods for analysis of barriers, factors and enablers of SM are presented in Table 12.1. A review of MCDM methods for concept, design and material selection for SM are presented in Table 12.2.

Among the various MCDM methods, the analytic hierarchy process (AHP) and technique for order preference by similarity to the ideal solution (TOPSIS) are widely utilized in many studies. AHP is used as it enables effective pairwise comparison to generate solutions. TOPSIS is practitioner-friendly as computations are simple but effective.

12.3 Research Agenda in Sustainable Manufacturing

Various research avenues in SM include

- Approaches for inducing sustainability in early product design stages (DFE, QFDE)
- Approaches for quantification of environmental impact (LCA, EIA)
- MCDM in SM (alternative selection in terms of materials, product design, process and so on)
- Societal aspects (societal LCA, CSR)
- SM indicators and performance measurement
- Lean extension to sustainable manufacturing
- Design strategies supporting SM
- SM integration with Industry 4.0

TABLE 12.1

Review on SM Barriers, Enablers and Factors

Reference	MCDM Technique	Focus
Garg et al. (2014)	DEMATEL	To evaluate drivers in deploying SM
Bhanot et al. (2015)	Statistical techniques	To analyze enablers and barriers of SM
Shankar et al. (2016)	AHP	To analyze drivers of advanced SM systems
Bhanot et al. (2017)	DEMATEL, ISM and SEM	To analyze the enablers and barriers of SM
Shankar et al. (2017)	DEMATEL	To analyze SM practices
Moktadir et al. (2018)	Graph theory and matrix approach	To prioritize drivers of SM practices
Malek and Desai (2019)	Best Worst Method	To prioritize SM barriers

TABLE 12.2

Review of MCDM Methods in SM Concept, Design and Material Selection

Reference	MCDM Technique	Focus
Zarandi et al. (2011)	Life cycle engineering, rule-based expert system	Material selection for sustainable product design
Vinodh and Girubha (2012a)	PROMETHEE	Sustainable concept selection
Vinodh and Girubha (2012b)	ELECTRE	Sustainable concept selection
Vinodh et al. (2013)	Modified fuzzy TOPSIS	Sustainable concept selection
Ocampo and Clark (2015)	AHP	Selection of SM initiatives
Vimal and Vinodh (2016)	LCA and ANP	Selection of SM processes
Khatri and Metri (2016)	SWOT–AHP	SM strategy selection
Singh et al. (2016)	AHP–VIKOR	SM strategy selection
Ribeiro et al. (2016)	Life cycle engineering principles	Selection of SM technologies
Bradley et al. (2016)	Life cycle costing with evolutionary algorithm	Sustainable material selection
Mousavi-Nasab and Sotoudeh-Anvari (2018)	A new MCDM approach	Sustainable material selection

Research centres in SM (both international and national) conduct dedicated research on various aspects of SM to contribute towards theory and practice of SM. Also, industry-oriented collaboration enables the practical validity of various SM theories. Also, preparations are being made to move towards the fourth industrial revolution (Industry 4.0).

12.4 Sustainable Manufacturing Portal

A portal for SM is an idea wherein a comprehensive database and software support can be planned. LCA is a potential tool wherein an LCA module along with an LCA database could be integrated with an SM portal to provide needful data pertaining to environmental impacts. The SM portal would focus on significant environmental impacts in the context of product design and manufacturing: carbon footprint, acidification, eutrophication and total energy consumption. The SM performance module could be integrated in the SM portal wherein SM performance indicators and dedicated models could be analyzed to provide insights to apply appropriate SM indicators as well as index quantification. Decision-making modules in view of material selection, alternative design selection, process selection and supplier selection could be integrated. Design for disassembly, recycling and environment modules could also be integrated with the portal to provide relevant information regarding product value gained from the 3Rs (reduce, reuse, recycle) and 6Rs (3Rs plus recovery, redesign and remanufacture). Also, life cycle cost analysis could be integrated.

12.5 Sustainable Manufacturing for Industry 4.0

Sustainable Industry 4.0 enables the amalgamation of business activities through interconnected elements and process innovation for manufacturing to be flexible, cost effective and eco-friendlier (de Sousa Jabbour et al., 2018a). Economic sustainability of the enterprise can be achieved using cloud-based computation based on the big data analytics platform thereby reducing the financial burden of infrastructure.

The real-time manufacturing cost tracking system based on integrated lean management with an RFID system identifies the unnecessary manufacturing cost and eliminates them (Bonilla et al., 2018). Implementation of lean management targets reduces the unnecessary cost involved in cost-intensive sustainable Industry 4.0 systems. The manufacturing process stability and its economic performance can be improved by real-time preventive maintenance of sensor systems using predictive analytics (de Sousa Jabbour et al., 2018b).

Sustainable manufacturing focuses on efficient use of energy sources. Steered with smart devices, Industry 4.0 systems have greater scope to minimize overproduction, production waste, materials movement and energy consumption on sustainability (Bonilla et al., 2018). Sustainable manufacturing necessitates the use of additive manufacturing to produce customized products with shorter lead times, reduced inventories and effective capacity utilization.

Sustainable Industry 4.0 ensures environmental protection through efficient use of resources in line with real-time data collection resulting in sustainable green practices (Bonilla et al., 2018). Industry 4.0 enables production process stability through digital management and avoids safety issues and operational risks to manufacturing workers.

Sustainable Industry 4.0 integrates lean principles that steady processes and minimize waste. Sustainable Industry 4.0 is expected to minimize operational costs by means of end-to-end digital amalgamation as against the high implementation cost (Kamble et al., 2018).

12.6 Summary

This chapter discussed research issues of sustainable manufacturing from the following perspectives: relation between lean and sustainable manufacturing, multi-criteria decision-making methods in sustainable manufacturing (barriers, enablers, factors prioritization and alternatives [concept, design, material, product] selection), research agenda in sustainable manufacturing, sustainable manufacturing portal, and sustainable manufacturing for Industry 4.0.

References

Bhanot, N., Rao, P.V., & Deshmukh, S.G. (2015). Enablers and barriers of sustainable manufacturing: Results from a survey of researchers and industry professionals. *Procedia CIRP, 29*, 562–567.

Bhanot, N., Rao, P.V., & Deshmukh, S.G. (2017). An integrated approach for analysing the enablers and barriers of sustainable manufacturing. *Journal of Cleaner Production, 142*, 4412–4439.

Bonilla, S.H., Silva, H.R., Terra da Silva, M., Franco Gonçalves, R., & Sacomano, J.B. (2018). Industry 4.0 and sustainability implications: A scenario-based analysis of the impacts and challenges. *Sustainability, 10*(10), 3740.

Bradley, R., Jawahir, I.S., Badurdeen, F., & Rouch, K. (2016). A framework for material selection in multi-generational components: Sustainable value creation for a circular economy. *Procedia CIRP, 48*, 370–375.

de Sousa Jabbour, A.B.L., Jabbour, C.J.C., Godinho Filho, M., & Roubaud, D. (2018a). Industry 4.0 and the circular economy: A proposed research agenda and original roadmap for sustainable operations. *Annals of Operations Research, 270*(1–2), 273–286.

de Sousa Jabbour, A.B.L., Jabbour, C.J.C., Foropon, C., & Godinho Filho, M. (2018b). When titans meet–can industry 4.0 revolutionise the environmentally-sustainable manufacturing wave? The role of critical success factors. *Technological Forecasting and Social Change, 132*, 18–25.

Garg, D., Luthra, S., & Haleem, A. (2014). An evaluation of drivers in implementing sustainable manufacturing in India: Using DEMATEL approach. *International Journal of Social, Behavioral, Educational, Economic, Business and Industrial Engineering, 8*(12), 3517–3522.

Kamble, S.S., Gunasekaran, A., & Gawankar, S.A. (2018). Sustainable Industry 4.0 framework: A systematic literature review identifying the current trends and future perspectives. *Process Safety and Environmental Protection, 117*, 408–425.

Khatri, J.K., & Metri, B. (2016). SWOT-AHP approach for sustainable manufacturing strategy selection: A case of Indian SME. *Global Business Review, 17*(5), 1211–1226.

Malek, J., & Desai, T.N. (2019). Prioritization of sustainable manufacturing barriers using Best Worst Method. *Journal of Cleaner Production, 226*, 589–600.

Moktadir, M.A., Rahman, T., Rahman, M.H., Ali, S.M., & Paul, S.K. (2018). Drivers to sustainable manufacturing practices and circular economy: A perspective of leather industries in Bangladesh. *Journal of Cleaner Production, 174*, 1366–1380.

Mousavi-Nasab, S.H., & Sotoudeh-Anvari, A. (2018). A new multi-criteria decision making approach for sustainable material selection problem: A critical study on rank reversal problem. *Journal of Cleaner Production, 182*, 466–484.

Ocampo, L., & Clark, E. (2015). An analytic hierarchy process (AHP) approach in the selection of sustainable manufacturing initiatives: A case in a semiconductor manufacturing firm in the Philippines. *International Journal of the Analytic Hierarchy Process, 7*(1), 32–49.

Ribeiro, I., Kaufmann, J., Schmidt, A., Peças, P., Henriques, E., & Götze, U. (2016). Fostering selection of sustainable manufacturing technologies: A case study involving product design, supply chain and life cycle performance. *Journal of Cleaner Production, 112*, 3306–3319.

Shankar, K.M., Kannan, D., & Kumar, P.U. (2017). Analyzing sustainable manufacturing practices: A case study in Indian context. *Journal of Cleaner Production, 164*, 1332–1343.

Shankar, K.M., Kumar, P.U., & Kannan, D. (2016). Analyzing the drivers of advanced sustainable manufacturing system using AHP approach. *Sustainability, 8*(8), 824.

Singh, S., Olugu, E.U., Musa, S.N., Mahat, A.B., & Wong, K.Y. (2016). Strategy selection for sustainable manufacturing with integrated AHP-VIKOR method under interval-valued fuzzy environment. *The International Journal of Advanced Manufacturing Technology, 84*(1–4), 547–563.

Varela, L., Araújo, A., Ávila, P., Castro, H., & Putnik, G. (2019). Evaluation of the relation between lean manufacturing, industry 4.0, and sustainability. *Sustainability, 11*(5), 1439.

Vimal, K.E.K., & Vinodh, S. (2016). LCA integrated ANP framework for selection of sustainable manufacturing processes. *Environmental Modeling and Assessment, 21*(4), 507–516.

Vinodh, S., & Girubha, R.J. (2012a). PROMETHEE based sustainable concept selection. *Applied Mathematical Modelling, 36*(11), 5301–5308.

Vinodh, S., & Girubha, R.J. (2012b). Sustainable concept selection using ELECTRE. *Clean Technologies and Environmental Policy, 14*(4), 651–656.

Vinodh, S., Mulanjur, G., & Thiagarajan, A. (2013). Sustainable concept selection using modified fuzzy TOPSIS: A case study. *International Journal of Sustainable Engineering, 6*(2), 109–116.

Vinodh, S., Ruben, R.B., & Asokan, P. (2016). Life cycle assessment integrated value stream mapping framework to ensure sustainable manufacturing: A case study. *Clean Technologies and Environmental Policy, 18*(1), 279–295.

Zarandi, M.H.F., Mansour, S., Hosseinijou, S.A., & Avazbeigi, M. (2011). A material selection methodology and expert system for sustainable product design. *The International Journal of Advanced Manufacturing Technology, 57*(9–12), 885–903.

Index

A

ABS, 71, 72, 104, 105
Acidification, 18, 23, 25, 51, 69, 70, 76, 80, 85, 86, 117
Acidification potential, 23, 76, 86
Active disassembly, 33
Additive manufacturing, 85, 87, 89, 90
Adequacy, 18
Air acidification, 51, 70
Air pollution, 60, 107, 108, 112
Analytic hierarchy process (AHP), 107, 116, 117
ANP, 117
Archetype, 27
Architecture, 33, 59
Assumptions, 16, 19, 26, 28, 78
Attraction, 55
Attributes, 32, 61, 62, 85
Automotive part, 71, 72

B

Background data, 17
Balance, 37, 74, 75, 115
Balanced score card, 62
Barriers, 116, 119
Base design, 36, 37
Baseline impact, 23
Battery recycling, 35
Benchmark, 16, 26, 27, 66, 69, 71, 73, 78
Benefit score, 50, 53–55
Best Worst Method, 116
Big data analytics, 118
Binder jetting, 88
Biodegradability, 93–95, 97, 98, 100, 101
Bottlenecks, 78
Boundaries, 16
Business associations, 56
Business authenticity, 56
Business ethics, 56

C

Carbon footprint, 46, 47, 51, 69, 70, 73, 117
Carcinogens, 80
Characteristics, 6, 53, 89
Characterization, 17, 18, 23–25, 27–29, 75, 78, 105
Classification, 17, 75
Cleaner production, 3
Climate change, 23, 37, 46, 80
CML, 23–24, 27–29, 86, 87
CML 2001, 23, 24, 27, 28
CO2PE, 87
Collaboration, 117
Common materials, 32
Communication, 43, 44
Community projects, 62
Company's image, 56
Company's reputation, 60
Competitive advantage, 37
Completeness, 19
Compliance, 3, 4, 39, 43, 60
Computer aided design (CAD), 39, 69
Concept design, 37
Conclusions, 18, 19
Conducive working environment, 61
Consistency, 19, 108, 109
Conventional manufacturing, 85, 90
Corporate social responsibility, 49, 50
Cost reduction, 3, 45, 55
Cradle to gate, 16, 17
Cradle to grave, 16, 37
Craft, 1, 6
Criterion, 50–53, 107–112
Customer requirements, 9, 11, 14, 93, 99
Customers, 38, 45, 54
Customer weight, 9–11, 94, 99–101

D

Database, 17, 24, 69, 70, 74, 78, 86, 87, 88, 117

Defects, 5
DEMATEL, 116
Description, 16, 23, 26, 29
Descriptive methods, 38
Design engineers, 39, 69
Design enhancement, 9
Design for additive manufacturing (DFAM), 89, 90
Design for disassembly, 31, 39, 117
Design for energy efficiency, 39
Design for environment, 31, 34, 36
Design for excellence (DfX), 39
Design for recycling, 31, 33
Design for separability, 35
Design freedom, 87, 89
Design improvement, 10, 11, 93, 96
Design principles, 89
Design rules, 89
Design strategies, 31, 39, 40, 89, 116
Design strategy, 36
Destructive disassembly, 33
Detail design, 37, 39
DFE matrix, 38
Dielectrics, 87
Disassembly directions, 32
Disassembly mechanism, 32
Disassembly points, 32
Disassembly time, 31, 32
Distance to target, 26
Drivers, 49, 55, 56, 89, 116
Durability, 36, 39

E

Ease of repair, 36
Eco design, 34, 36, 37, 39
Eco efficiency, 37
Eco FMEA, 38
Eco friendlier product, 13, 34
Eco friendliness, 104
Eco friendly, 31, 36, 37, 69
Eco friendly product design, 31, 37
Eco friendly products, 69
Eco indicator, 25–29, 75, 79, 80, 87, 88, 95, 96, 102
Eco innovative product design, 39
Ecoinvent, 24, 86–88
Ecological balance, 115
Economic development, 54
Economies of scale, 1
Economy, 1, 2, 5, 59–63, 67
Eco redesign, 38
Ecosystem, 25, 27, 28, 80, 105
Ecosystem quality, 27, 80, 105
Ecotoxicity, 76, 80, 88
Eco value analysis, 38
Efficiency optimization, 86
Egalitarian, 27, 28
EIA, 116
ELECTRE, 117
Electric discharge machining, 86
Electro chemical machining, 86
Elementary flow, 18
Embodiment design, 37
Emergy, 88
Emission reduction, 35, 54
Emissions, 4, 5, 17, 18, 24, 26, 37, 45, 46, 56, 78, 79, 85, 87, 115
Employee relations, 56
Employee satisfaction, 62
Employee turn over ratio, 61
Employment compensation, 61
Employment opportunity, 107, 108, 110
Employment rate, 107, 108, 110
EMS audit, 44
Enablers, 61, 62, 66, 116, 119
End of life, 16–17, 31, 34–36, 49, 53, 75, 88
Endorsements, 19
Endpoint, 24, 26, 27, 28, 29
Energy aspects, 85
Energy consumed, 51, 70
Energy consumption rate, 88
Energy efficiency, 36, 39, 60, 90
Energy utilization, 4, 85
Engineering metrics, 9–11, 14, 93–98, 100, 101
Enhanced image, 55
Environmental challenges, 13
Environmental concerns, 13, 34, 37, 54
Environmental cost, 19–21, 38, 103
Environmental declaration, 16
Environmental FMEA, 38
Environmental indicators, 70
Environmental legislation, 60
Environmental management system, 43
Environmental performance, 34, 39, 45, 49, 51–53, 55

Environmental performance index, 49, 51–53, 55
Environmental policy, 13, 26, 44
Environmental protection, 3, 118
Environmental QFD, 38
Environmental regulations, 37
Ethics and transparency, 61
Euclidean distance, 66
Eutrophication, 18, 23, 25, 51, 69, 70, 72, 76, 80, 85
Evacuation, 51
Experimental, 85, 87
Extended producer responsibility (EPR), 13, 14
Extent of government support, 61
External trigger, 33

F

Failure mode and effect analysis, 38
Fasteners, 31–33
Financial ratios, 56
Flows, 23–25, 46, 73–75, 78, 79
Fluke power analyzer, 85
Foreground data, 17
Functional unit, 16, 26, 74, 75
Fused deposition modelling (FDM), 87–89, 102, 104
Fused layered manufacturing (FLM), 88
Fuzzy based, 116
Fuzzy Logic, 59, 64, 66, 67
Fuzzy performance importance index, 66

G

GaBi, 69, 72–76, 80, 86–88
Gate to gate, 16, 17
Gate to grave, 16, 17
GHG emissions, 45, 46, 85
Global matrix, 112
Global score, 11, 99–101
Global warming potential, 70, 76, 86
Global weight, 112, 113
Goal alignment, 56
Government regulations, 56
Government rules, 61
Graph theory and matrix approach, 116
Greenhouse effect, 25
Green manufacturing, 2
Green production, 3, 54
Green QFD, 38
Grey theory based, 116
Grouping, 17, 18, 31

H

Hazardous content, 51
Hazardous material ratio, 60
Health issues, 61, 86
Hierarchist, 27, 28
Housekeeping, 4
House of ecology, 38
Human health, 25, 27, 28, 80, 105
Human toxicity, 23, 76

I

IMPACT, 2002+, 87
Impact assessment, 15, 17, 19, 21, 23, 24, 29, 74, 75, 78, 80, 86, 87, 102
Impact category, 18, 23, 24
Implementation cost, 107, 108, 110, 118
Improvement effect, 11, 14, 99–101
Improvement rate, 10, 11, 14, 96, 99–101
Individualist, 27, 28
Industry, 4.0 116, 118
Inferences, 19, 26, 85, 86, 88
Influencing factor, 49, 50, 62
Input-output-impact, 75
Interpretation, 15, 18, 19, 21, 74–76, 78, 80, 102
Inventory, 5, 15, 17, 19, 21, 74, 75, 78, 80, 87, 88, 102, 115
Inventory analysis, 15, 17, 21, 74, 75, 80, 102
Investment, 60
Investors, 56
Ionizing radiation, 76
ISM, 116
ISO 14000, 45
ISO 14040, 15
ISO 14044, 15

J

Job enhancement, 56
Job opportunities, 61
Job satisfaction level, 61

K

Kano model, 39

L

Labour issues, 61
Landfill quantity, 52
Laser engineered net shaping (LENS), 87, 88
Layer thickness, 89
Level data, 50
Life cycle assessment, 15–17, 19, 21, 43, 45, 69, 86, 90, 102, 113, 115
Life cycle costing, 19–21, 73, 74, 102, 103, 113, 117
Life cycle environmental cost analysis, 38
Life cycle impact assessment (LCIA), 23, 26, 28, 29, 75, 80, 86, 88
Life cycle inventory, 15, 17, 74, 78
Life cycle phases, 4, 15, 17, 37, 50, 54, 56, 76, 78, 104
Life cycle planning, 39
Life extension, 35
Lightweight products, 87
Limitations, 16, 19
Liquid waste, 60

M

Machining performance, 60
Macro social performance, 61
Manufacturing cost, 60, 118
Manufacturing process, 69, 70, 72, 73, 118
Market dynamism, 1
Market price, 20, 21, 103, 129
Market share, 5, 45, 60
Mass manufacturing, 1, 6
Mass production, 1
Material choice, 51
Material extraction, 16, 17, 35, 49, 51
Material handling, 52
Material intensity, 62
Material substitution, 4, 35
Material utilization, 35, 60
Mathematical modelling, 87
Midpoint, 18, 23, 24, 26–29
Millipoint, 27
Mineral and energy resources, 60
Modelling, 23, 74, 76, 78, 87, 102
Monetization, 27, 29
Monetization factor, 29
Motivation, 55
Multi criteria decision making, 59, 107, 113, 115, 119
Multi grade fuzzy, 59, 63, 67

N

Net product gain, 21, 102, 104
Network based, 115
Network complexity, 52
Network diagram, 79
New product design, 4
Ninth waste, 5, 115
Noise level, 53
Non destructive disassembly, 33
Non-renewable material ratio, 60
Normalization, 17, 18, 23, 24, 27–29, 78
Normalization factor, 23, 24

O

Occupational health, 61
Operational efficiency, 87
Organisation for Economic Cooperation and Development (OECD), 13
Orientation, 85, 88, 89
Outranking, 115
Overproduction, 5, 118
Ozone depletion potential, 76, 87
Ozone layer depletion, 25

P

Packing method, 52
Pairwise comparison, 107, 116
Panel, 27, 29
Paradigm, 2
Particulate matter, 76
Part orientation, 89
PAS 2050, 45, 46, 47
People, 1, 2, 62
Photochemical ozone creation potential, 76

Physical recycling, 35
Planet, 1, 2, 62
Plastics recycling, 35
Pollutants, 5, 62, 115
Pollution abatement, 3
Pollution prevention, 3
Portal, 117, 119
Potential benefits, 13
Priority, 14, 18, 112, 113
Process improvement, 61
Processing, 5, 16–17, 20, 49, 51, 75, 102, 104
Process map, 46, 74
Process optimization, 4
Process selection, 69, 70, 115, 117
Procurement cost, 20, 21, 103
Product architecture, 33
Product complexity, 1
Product cost, 19, 20, 39, 103
Product development, 9, 16, 37
Product effectiveness, 19, 103
Product EoL, 60
Product gain, 20, 21, 102–104
Production cost, 35, 85
Production efficiency, 61
Production methods, 52
Productivity, 1, 3–4, 45, 60, 88
Product life cycle, 9, 15, 19–21, 26, 34, 35, 46, 50–54, 56, 71, 78, 93, 102, 103
Product life cycle cost, 19–21, 102, 103
Product recovery, 31, 46
Product service, 31
Product structure, 32
Product sustainability, 49, 55, 72
Product value, 20, 21, 24, 103, 117
Profitability, 1, 45, 60, 62
PROMETHEE, 117

Q

QFD for environment (QFDE), 38, 116
Quality aspect, 62
Quality function deployment, 9, 13, 38, 93, 113
Quality of life, 54

R

Radio frequency identification (RFID), 115, 118
Ranking, 18, 112–113
Rating, 9, 10, 63, 65, 66
Raw material extraction, 35, 51
Raw score, 10, 93–96
ReCiPe, 28, 29, 87, 88
Reconditioning, 53
Recovery, 2, 31, 46, 47, 53, 118
Recovery options, 53
Recyclability, 33–35, 39
Recyclable material, 33, 60
Recycle, 2, 117
Recycling characteristics, 53
Recycling potential, 36, 93–95, 97, 98, 100, 101
Redesign, 2, 38, 118
Regulations, 3, 4, 37, 39, 56, 60, 61
Relational strength, 10, 11, 93
Relative weight, 10, 94–98
Relays, 93
Reliability, 19, 93, 94, 99–101
Remanufacture, 2, 35, 118
Remanufacturing characteristics, 53
Remanufacturing cost, 20, 36, 103
Renewable energy ratio, 60
Repair alternatives, 53
Reputation, 3, 4, 55, 56, 60, 76
Requirements matrix, 38
Resource consumption, 4, 35, 60, 61, 85
Resources depletion, 27
Resource utilization, 4, 35, 45, 115
Respiratory inorganics, 80
Respiratory organics, 80
Reuse, 2, 53, 117
Reused material ratio, 60
Reuse possibilities, 53
Reverse manufacturing, 19–21
Risk score, 50, 53–55

S

Safety issues, 61, 118
Safety risk, 61
Salary and benefits, 60
Security, 61
Selective disassembly, 33
Selective laser sintering (SLS), 88
SEM, 116
Sensitivity, 19, 78
SimaPro module, 76, 78, 113

Skill level of worker, 61
Smart materials, 33
Smog, 25
SM portal, 117
Social cohesion, 61
Societal LCA, 116
Society, 50, 54, 59, 61–63, 67
Soil pollution, 60
Solid waste, 60
Stakeholder participation, 61
Stakeholders, 45, 50, 54
Stakeholder value, 1
Stereolithography apparatus (SLA), 87
Strategic advantage, 56
Subjective weighting, 59
Subsidy, 56
Substances, 24, 25, 27
Supplier selection, 115, 117
Supply chain, 52, 78
Sustainability analysis, 69, 80
Sustainability enablers, 61, 62, 66
Sustainability evaluation, 62, 67, 85
Sustainability index, 62, 65–67
Sustainability indicators, 59–61, 63, 65, 67
Sustainability levels, 66
Sustainability report, 71–73
Sustainability Xpress, 69–71
Sustainable AM, 87
Sustainable product design, 117
System boundaries, 16, 73–74

T

Technique for order preference by similarity to the ideal solution (TOPSIS), 116, 117
Technology selection, 115
3D Printers, 88, 89
3Rs, 6Rs, 2, 117
Time horizon, 24, 29
Total disassembly, 33
Toxic fumes, 51
Toxic substances, 25
Toyota production system, 2
TRACI, 87
Trading opportunities, 60
Training hours/employee, 61
Transport, 5, 36, 46
Transportation, 16, 17, 49, 52, 69, 70, 73
Triple bottom line, 1, 2, 6, 54, 59, 85
TRIZ, 39

U

Ultrasonic machining, 86
Uncertainty, 23, 24, 28, 34
Unconventional manufacturing, 86, 90

V

Value analysis, 38
Value creation, 2
Value stream mapping (VSM), 115
Vendors, 54
VIKOR, 117
Visual map, 73
Voice of customer, 9, 93, 94, 100, 101

W

Waiting, 5
Waste management, 13, 61
Waste material ratio, 60
Waste source, 35
Waste water ratio, 60
Water eutrophication, 51, 70, 72
Water use, 60
Weighting, 17, 18, 23–29, 59, 78
Weighting method, 23, 25, 27
Weighting Triangle, 27, 29
Welding process, 79–82
Wood recycling, 35
Worker health, 85
Worker safety, 85
Workforce training, 61
Workplace environment, 61
World Business Council for Sustainable Development (WBCSD), 54

Printed in the United States
By Bookmasters